PRACTICAL GUIDE TO RESPIRATOR USAGE IN INDUSTRY

PRACTICAL GUIDE TO RESPIRATOR USAGE IN INDUSTRY

Second Edition

Gyan S. Rajhans
Bhawani P. Pathak

An imprint of Elsevier Science

Amsterdam   London   New York   Oxford   Paris   Tokyo
Boston   San Diego   San Francisco   Singapore   Sydney

Butterworth–Heinemann is an imprint of Elsevier Science.

Library of Congress Cataloging-in-Publication Data

Rajhans, Gyan S.
  Practical guide to respirator usage in industry / Gyan S. Rajhans, Bhawani P. Pathak.—
2nd ed.
    p. cm.
  Includes bibliographical references and index.
  ISBN 0–7506–7435–0 (alk. paper)
  1. Breathing apparatus. 2. Respiratory organs—Diseases—
  Prevention. 3. Gases, Asphyxiating and poisonous—Safety
  measures. 4. Industrial safety—Equipment and supplies.
  I. Pathak, Bhawani. II. Title.
  T55.G3 R35   2002
  604.7—dc21   2001056550

British Library Cataloguing-in-Publication Data
A catalogue record for this book is available from the British Library.

The publisher offers special discounts on bulk orders of this book.
For information, please contact:

Manager of Special Sales
Elsevier Science
225 Wildwood Avenue
Woburn, MA 01801-2041
Tel: 781-904-2500
Fax: 781-904-2620

For information on all Butterworth–Heinemann publications available, contact
our World Wide Web home page at: http://www.bh.com

10 9 8 7 6 5 4 3 2 1

Printed in the United States of America

Contents

Preface

The first edition of *Practical Guide to Respirator Usage in Industry*, published in 1985, was well received by the manufacturers and users of respirators. One of the reviewers, R.M. Howie, in the July 1986 issue of *Annals of Occupational Hygiene*, declared: "In short, this is an excellent book." Another reviewer, D.A. Deeds, in the February 1986 issue of *American Industrial Hygiene Journal*, said: "The respirator program in Chapter 7 is very complete and detailed. Overall the book is a good guide for a respirator program."

After an interval of about 16 years, we believe that the respirator manufacturers, the National Institute for Occupational Safety and Health (NIOSH), and academicians in industrialized countries have made tremendous progress toward the "Research Needs" identified in Chapter 9 of the first edition, hence the need for the second edition. We particularly identified the lack of data on the wearer component of respiratory protection. Until 1985, only limited research had been conducted either on comfort characteristics of a respirator or on the psychological aspects of wearing a respirator. It is gratifying to note that numerous research papers have since been published dealing with these problems; however, another important research need pointed out in the first edition has not been adequately fulfilled. To the best of our knowledge, little research has been published regarding respiratory protection of female workers, who are fast becoming a significant percentage of the total workforce throughout the world. We encourage researchers to concentrate on this growing need.

The U.S. OSHA 29CFR 1910.134 Respirator Regulations with their appendices were published in January 1998. We find them to be the most comprehensive respiratory protection equipment (RPE) regulations promulgated in industrialized nations. While referring to this regulation in various chapters of this book, we have provided a few case studies in Chapter 8 to demonstrate noncompliance with OSHA regulations.

Chapter 8 also has added several new case studies because, in our opinion, we can learn a lot about the problems and issues associated with respirator usage by reviewing case studies. These case studies also reveal that engineering and administrative controls have failed thus far to eliminate respiratory hazards in most workplaces. Thus, RPE continues to be an important component in many respiratory exposure control plans. Respiratory hazard evaluation; RPE selection and fitting, use, evaluation, and maintenance; and medical supervision of the wearers continue to be important aspects of effective respiratory protection.

We have, therefore, considerably updated Chapters 1 through 7 to reflect the progress made in these areas of research related to RPE. In order to make this edition user-friendly and to assist those unfamiliar with the technical terms and acronyms of respiratory protection, we have provided a comprehensive list of definitions at the beginning of the book. The lack of this information in the first edition was a matter of concern for some reviewers. Regulatory updates have been provided where required in each chapter, while maintaining the approach of the first edition. Most of the pictures and illustrations have also been replaced with better photographs and schematic diagrams, where needed. A list of Website addresses containing information on regulatory updates, respiratory performance, respirator classification and certification, and respirator selection and use is provided in the Appendix.

Inhalation is the major route of entry of toxic materials into the human body. As mentioned in the first edition, respiratory protection is permitted when, for whatever reason, engineering controls fail to provide adequate protection. This book is a guide for employers and employees who use respiratory protection; it is not an attempt to provide a detailed look at all the published literature on respiratory protection. This guide is for plant managers, safety officers, plant nurses, industrial hygienists, and others, who have to develop and implement a complete respiratory protection program.

As noted in the first edition and emphasized in this edition, total respiratory protection programs will not be effective until they include occupational health and safety professionals, senior managers and supervisors, coworkers, and, most important, the wearers themselves. These individuals should work as a team to develop strategies to improve wearers' abilities to communicate, to minimize their visual problems, to maximize the comfort of RPE, and to decrease psychological problems that compound the difficulties of wearing respirators. Senior management also has to demonstrate commitment and support to worker health and safety because this concern drives workers to use RPE properly.

It is our hope that this practical guide, along with the case studies, will aid users in establishing a proper and effective respiratory protection program in accordance with the current legislation and, in turn, will protect the health of workers. In case it is misunderstood, rather than using "he or she" throughout the book, we have opted to use the universal "he."

The authors gratefully acknowledge the contribution of Kelly Donovan of Butterworth Publishers, whose diligent formatting, editing, and proofreading made this book possible. We wish to thank Lewis Publishers; CRC Press, Boca Raton, Florida; 3M Canada; and North Safety Products for permission to use their schematic diagrams and photographs. Special acknowledgement and appreciation are given to the Canadian Centre for Occupational Health and Safety (CCOHS), Hamilton, Canada; and Resource Environmental Associates, Toronto, Canada, for facilitating completion of this book.

G.S.R.
B.P.P.

Definitions

Acute effects: Effects that occur rapidly after exposure and are of short duration.

Administrative control measures: Methods of controlling worker exposure by limiting contact time with the contaminant.

Aerodynamic diameter: The diameter of a unit-density spherical particle with the same settling velocity in air as the particle in question of arbitrary shape and density.

Aerosol: Particles—solid or liquid—that remain suspended in air for a period. Aerosols include mists, smokes, fumes, and dusts.

AIHA: American Industrial Hygiene Association.

Air pollution: Presence of substances in the atmosphere resulting either from human activity or natural processes in sufficient concentration for a sufficient time and under circumstances such as to interfere with the comfort, health, or well-being of persons or the environment.

Air-purifying respirator (APR): A respirator employing an air-purifying element to remove specific air contaminants.

Allergen: Substances that produce an allergic reaction.

American Conference of Governmental Industrial Hygienists (ACGIH): Publishes threshold limit values (TLVs) for chemical substances and physical agents and biological exposure indices (BEIs).

American National Standards Institute (ANSI): Publishes consensus standards.

Approved: Tested and certified by the National Institute for Occupational Safety and Health (NIOSH).

Assigned protection factor (APF): The minimum expected workplace level of respiratory protection that would be provided by a properly functioning respirator or class of respirators.

Atmosphere-supplying respirator: A respirator that supplies the user with breathing air from a source independent of the ambient atmosphere and includes supplied-air respirators (SARs) and self-contained breathing apparatus (SCBA).

Biological exposure indices (BEIs): The guidance values published by ACGIH for assessing biological monitoring results. The BEI generally indicates a concentration below which nearly all workers should not experience adverse health effects (e.g., BEI for lead in blood is $30 \mu g/100 \, ml$).

Biological monitoring: Analysis of exhaled air, a biological fluid (e.g., urine, blood, perspiration), or a body component (e.g., hair, nails).

Canister (air-purifying): A container filled with sorbents and catalysts that remove gases and vapors from air drawn through the unit. The canister may also contain an aerosol (particulate) filter to remove solid and liquid particles.

Carcinogen: An agent—physical, chemical, or biological—that can act on living tissue to cause cancer.

Cartridge: A small container filled with air-purifying media.

Certified: See *Approved.*

CGA: Compressed Gas Association.

Chronic effects: Effects that develop slowly and have long duration. They are often, but not always, irreversible. Some chronic effects appear a long time (several years) after exposure.

Compressed-breathing gas: Oxygen or air stored in a compressed state and supplied to the wearer in gaseous form.

Confined space: An enclosure such as a storage tank, process vessel, boiler, silo, tank car, pipeline, tube, duct, sewer, underground utility vault, tunnel, or pit that has limited means of getting out and poor natural ventilation and that may contain hazardous contaminants or be oxygen deficient.

Contaminant: A harmful, irritating, or nuisance material that is foreign to the normal atmosphere.

Demand respirator: An atmosphere-supplying respirator that admits breathing air to the facepiece only when a negative pressure is created inside the facepiece by inhalation.

Dioctyl phthalate (DOP): An oily particle commonly used for quantitative fit testing.

Disposable respirator: Also known as *maintenance-free respirators.* A respirator that is designed to be discarded after the end of its recommended period of use, after excessive resistance or physical damage, or when odor breakthrough or end-of-service-life renders it unsuitable for further use.

Dust: A mechanically produced solid particle (e.g., crushing, drilling, grinding, sweeping, or handling of solid materials).

Emergency respirator use: The wearing of a respirator for escape or rescue from an uncontrolled significant release of an airborne contaminant.

End-of-service-life indicator (ESLI): A system that warns the wearer of a respirator of the approach of the end of adequate protection (e.g., that the sorbent is approaching saturation or is no longer effective).

Exhalation valve: A device that allows exhaled air to leave a respiratory device and prevents outside air from entering through the valve.

Facepiece: That portion of a respirator that covers the wearer's nose, mouth, and/or eyes. Designed to make a gas-tight or dust-tight fit with the face, it includes the headbands, exhalation valve(s), and connections for an air-purifying device.

Filter: A medium used in respirators to remove solid or liquid particles from the airstream entering the respiratory enclosure.

Fit-check (negative/positive): A procedure used to determine if the respirator is properly adjusted by blocking the intake port(s), the exhaust port(s), and inhaling and exhaling, respectively. This procedure should be done each time a respirator is used.

Fit factor: A measure of the effectiveness of the facepiece to face seal determined by a quantitative fit test. It is the ratio of test agent concentration outside a respirator to the test agent concentration inside the respirator.

Fit test: The use of a protocol to qualitatively or quantitatively evaluate the fit of a respirator on an individual. See *Qualitative fit test* and *Quantitative fit test*.

Forced expiratory volume in one second (FEV$_1$): Volume of air that can be forcibly expelled during the first second of expiration.

Forced vital capacity (FVC): Maximal volume of air that can be exhaled forcefully after a maximal inspiration.

Fume: Solid particles generated by condensation from the gaseous state, generally after volatilization from a melted substance (e.g., welding) and often accompanied by a chemical reaction such as oxidation. Gases and vapors are not fumes.

Gas: A substance that is in the gaseous state at room temperature and pressure.

Hazardous atmosphere: Any atmosphere that is oxygen-deficient or that contains a toxic or disease-producing contaminant. The atmosphere may or may not be IDLH (see *IDLH atmosphere*).

High-efficiency particulate air filter (HEPA): A filter designed to remove 99.97% of specific-type particle material (DOP of 0.3 micrometers in diameter) from air. The equivalent NIOSH 42 CFR Part 84 particulate filters are the N100, R100, and P100 filters.

IDLH atmosphere: An atmosphere that is "immediately dangerous to life or health" (IDLH). An IDLH atmosphere poses an immediate hazard to life, such as being oxygen-deficient (containing less than 19.5% oxygen), or produces an irreversible debilitating effect on health.

Inhalation valve: A device that allows respirable air to enter the facepiece and prevents exhaled air from leaving the facepiece through the intake opening.

Loose-fitting facepiece: A respiratory inlet covering designed to form a partial seal with the face.

Mist: An aerosol consisting of liquid particles generated by condensation of a substance from the gaseous to the liquid state.

MSHA: Mine Safety and Health Administration.

NaCl: Sodium chloride; a solid particle commonly used for quantitative fit testing.

National Institute for Occupational Safety and Health (NIOSH): The U.S. federal agency that tests, approves, and certifies respiratory protection equipment.

Negative-pressure respirator: A respirator in which the air pressure inside the facepiece is negative during inhalation with respect to the ambient air pressure outside the respirator.

NFPA: National Fire Protection Association.

Nuisance dust: Innocuous dust not causing a serious pathological condition.

Occupational Safety and Health Administration (OSHA): The U.S. federal agency that sets the minimum requirements for respirator use.

Odor threshold: The lowest concentration of an air contaminant that can be detected by smell.

Oxygen deficiency: A sea-level (partial pressure of less than 148 millimeters of mercury) concentration of oxygen in the ambient air of less than 19.5% by volume.

Particulate matter: A suspension of fine solid or liquid particles in air, such as dust, fog, fume, mist, smoke, or sprays. Particulate matter suspended in air is commonly known as an *aerosol.*

Permissible exposure limit (PEL): Uppermost limit that a worker can be legally exposed to, as established by the Occupational Safety and Health Administration (OSHA).

Positive-pressure respirator: A respirator in which the pressure inside the respiratory inlet covering exceeds the ambient air pressure outside the respirator.

Powered air-purifying respirator (PAPR): An air-purifying respirator that uses a blower to force the ambient air through air-purifying elements to the respiratory inlet covering.

Pressure-demand respirator: A positive-pressure atmosphere-supplying respirator that admits breathing air to the facepiece when the positive pressure is reduced inside the facepiece by inhalation.

Protection factor (PF): The overall protection afforded by a certain type of respirator as defined by the ratio of the concentration of contaminant outside a face-mask to that inside the equipment under conditions of use. For example, if a half-mask respirator has a protection factor of 10, it may be used for protection in atmospheres with a contaminant concentration up to 10 times the permissible exposure limit (PEL).

Pulmonary function test: Tests requiring use of an approved spirometer including forced vital capacity (FVC), the maximum amount of air that can be expired from the lung after full inhalation, and forced expiratory volume after one second (FEV1), the amount of air forcibly expired in one second after full inhalation.

Qualitative fit test (QLFT): A test procedure to determine the effectiveness of the seal between the face-mask and the wearer's face. This pass/fail means of testing that relies on the subject's sensory response to detect the challenge agent is subjective and could pass the user without truly knowing if an adequate fit exists.

Quantitative fit test (QNFT): The measurement of the effectiveness of a respirator seal in a test atmosphere inside a booth. This test is performed by numerically measuring the amount of leakage into the respirator (see *Fit factor*).

Resistance: Opposition of the flow of air, as through a canister, cartridge, or particulate filter.

Respirator: A device designed to protect the wearer from inhalation of harmful atmospheres.

Respiratory inlet covering: A portion of a respirator that forms the protective barrier between the wearer's respiratory tract and an air-purifying device or breathing air source or both. It may be a facepiece, helmet, suit, or a mouthpiece respirator with nose clamp.

RPE: Respiratory protection equipment.

Self-contained breathing apparatus (SCBA): A respirator designed to provide the wearer with clean air independent of the contaminated surrounding air. The wearer carries a supply of approved compressed air contained in a gas cylinder. SCBA units are generally restricted to types equipped with pressure-demand regulators that maintain positive pressure in a full face-mask.

Service life: The length of time during which a respirator, filter, sorbent, or other respiratory equipment provides adequate protection to the wearer.

Simulated protection factor (SPF): Surrogate measure of the workplace protection provided by a respirator determined in a laboratory simulation that has been shown to have a stated correlation to workplace protection factors.

Smoke: Aerosols, gases, and vapors resulting from incomplete combustion.

Sorbent: A material contained in air-purifying respirators that removes toxic gases and vapors from the inhaled air.

Supplied-air respirator (SAR): A hose-mask respirator equipped with a facepiece, breathing tube, safety harness, and safety line. Air is supplied through a hose connected to a compressed-air cylinder or air compressor not designed to be carried by the user.

Test subject: A person wearing a respirator for quantitative fit testing.

Threshold limit value (TLV): A list published yearly by the American Conference of Governmental Industrial Hygienists (ACGIH) as a guide for exposure concentrations that a healthy individual normally can tolerate for eight hours a day, five days a week, without harmful effects. Airborne particulate concentrations are generally listed as milligrams per cubic meter of air (mg/m^3). Gaseous concentrations are listed as parts per million (ppm) by volume.

Tight-fitting facepiece: A respiratory inlet covering that forms a complete seal with the face.

Vapor: The gaseous state of a substance that is solid or liquid at room temperature and pressure.

Workplace protection factor (WPF): A protection factor provided in the workplace, under the conditions of that workplace, by a properly functioning respirator that is correctly selected, fit tested, worn, and used.

1

Respiratory Hazards and Evaluation

The type and extent of respiratory hazards present in the modern industrial environment dictate the selection of proper respirators. Some occupational activities and/or environments require the extra protection of respiratory equipment specifically designed to protect against hazards that may enter the body through the nose and mouth when a person breathes. Like clean air, many of these hazards are invisible and odorless. Breathing (or respiratory) hazards include dusts, fumes, mists; gases and vapors; and oxygen-deficient atmospheres. Knowing the characteristics of each hazard helps us understand why respiratory protection is so important. Because several good texts are available on this subject, an overview only of respiratory hazards encountered in industrial workplaces and the various methods of evaluating them are provided here.

The main categories of hazards are shown in Figure 1.1. In the case of respirators, we are mainly concerned with chemical hazards.

Any chemical that has the capacity to produce injury or illness when taken into the body is called a *toxic agent*. In order for an agent to develop toxic activities, it has to enter the body, be distributed to the body, and react with components of the body.

The human body has the capacity to eliminate ingested or inhaled noxious agents by vomiting, coughing, and excretion. Healthy skin, for example, will not absorb water-soluble substances; therefore, only substances that dissolve in body fluids can be toxic. Lead sulfide, for example, cannot be dissolved either in water or in body fluids, and is therefore not toxic, whereas other lead compounds, which can be dissolved, have a toxic action on the body. Another example is the iron compounds. Taken in large amounts, they are toxic; however, because they are corrosive and locally irritating in the stomach, they will be expelled from the stomach by vomiting before they can be absorbed.

In summary, in order to become toxic, an agent must be absorbed into the body (except some particulate materials, e.g., asbestos and silica, see the section on Pariculate Material), be dissolved in body fluids, and interfere with metabolism. For example, all lead compounds are inherently toxic; however, lead sulfide will not dissolve in body fluids unless taken orally. It is then transformed into lead

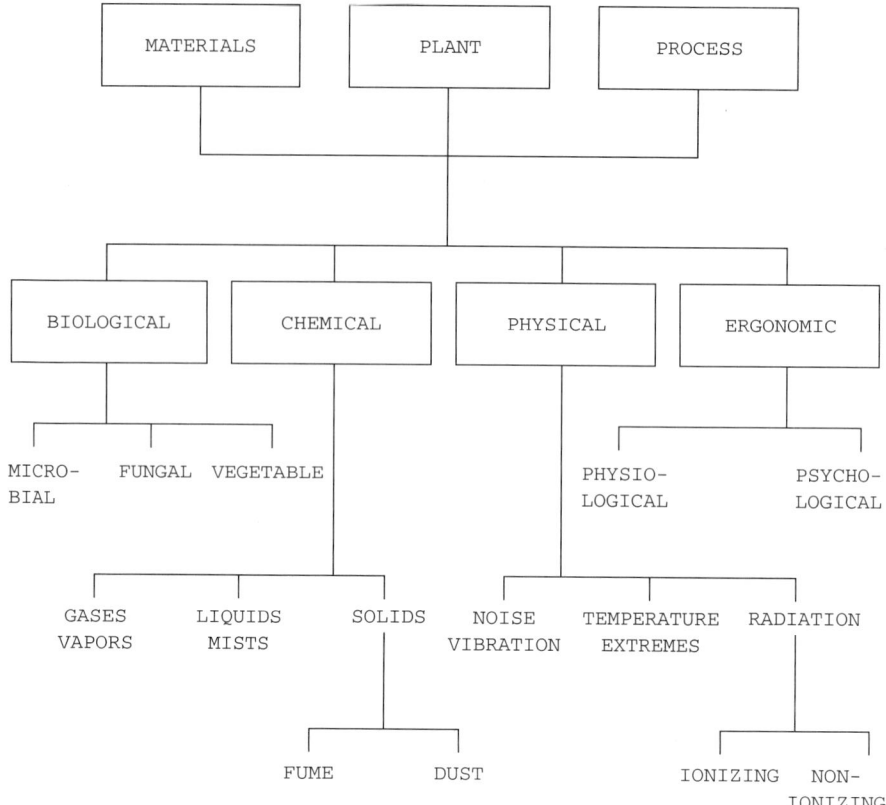

Figure 1.1 Occupational health hazards.

chloride by hydrochloric acid in stomach juices; therefore, its toxicity under industrial conditions is low.

A hazardous chemical, however, can be defined as carrying with it the high probability that its use will cause biological effects in the body. Any given chemical agent may have intrinsic toxicity, but it may not actually represent a health hazard unless the amount or the concentration that enters the body is sufficient to cause injury. Examples are trichloroethylene and perchloroethylene: both are chlorinated hydrocarbons widely used in industry as solvents and both produce the same toxic symptoms; however, the vapor pressure of trichloroethylene is 70 mmHg at 25 °C, whereas the vapor pressure of perchloroethylene is only 23 mmHg at 25 °C. This means that at the same room temperature, much more trichloroethylene will evaporate and its concentration in the inhaled air will be much higher than would be the concentration of perchloroethylene vapors. The toxicity of both compounds is the same; however, the hazard associated with perchloroethylene is lower because the concentration of its vapor in the inhaled air will be, under equal conditions, much lower.

ENTRY ROUTES FOR TOXIC SUBSTANCES

There are three routes by which chemical agents can enter the body:

1. By inhalation through the respiratory tract
2. By ingestion into the digestive tract through eating or swallowing
3. By contact with, or absorption through, the skin

As far as the use of respirators is concerned, the most important route of entry is by inhalation. Hence, this mechanism is discussed in detail.

The lungs have an extremely large surface area (approximately $140\,m^2$, including the alveolar surface and the blood capillary network surface), which provides a contact for the toxic materials. The large volume of air that is taken into the lungs on a daily basis, and the rapid rate of absorption from the air in the alveoli into the

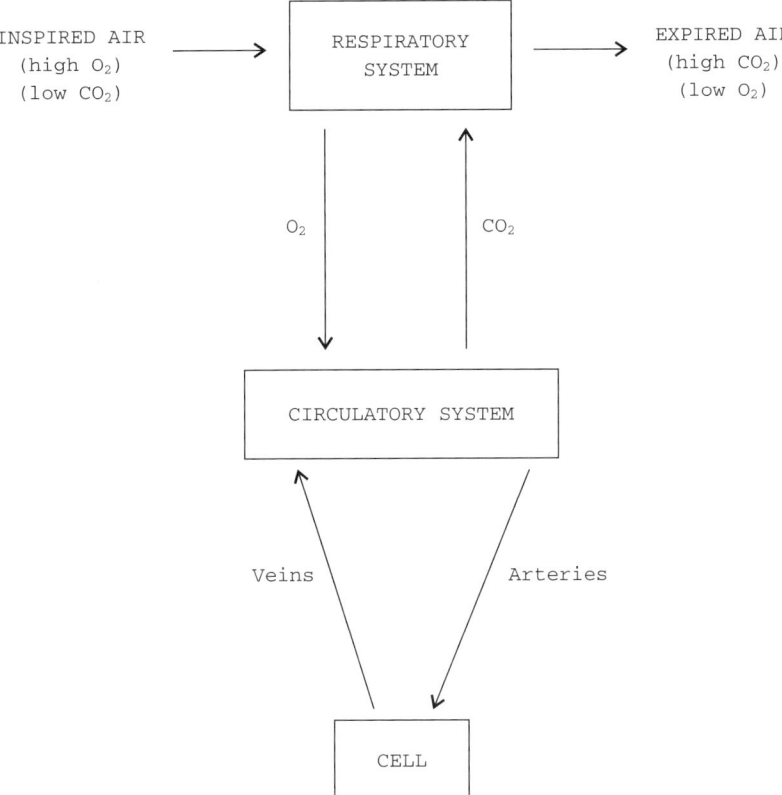

Figure 1.2 Role of the respiratory system in the transportation of oxygen and carbon dioxide.

bloodstream, make the respiratory tract the primary route for toxic materials to enter the body.

The primary function of the respiratory system is to provide a means for exchanging oxygen and carbon dioxide (Figure 1.2). Inhaled air enters the respiratory tract and is transported to the alveoli of the lungs. The alveoli are in close contact with blood capillaries and, because of concentration differences, oxygen diffuses from the alveoli to blood and carbon dioxide diffuses from blood in the capillaries to the alveoli. Once oxygen reaches blood, it combines with hemoglobin in the blood and is transported through the bloodstream to the cells. In the cells, oxygen is released from blood to provide food, which is necessary for cell life. In return, cells produce carbon dioxide as a waste byproduct. This carbon dioxide is released into the bloodstream and transported back to the lungs, where it diffuses into the lungs and is exhaled out of the body. The large surface area and the minute separation between the alveoli and the capillary system make the lungs an efficient organ for absorbing toxicants. These toxicants can be in the form of gases, solids, or liquid aerosols. Many toxicants are capable of being transported from the lungs to other parts of the body and exerting their toxic effects. Toxicants that exert direct action on the lungs can, and do, cause effects on the gas-exchange process, which can have acute and/or chronic effects on one's health.

The lungs are also involved in excreting toxicants that have entered the body through the skin or have been ingested. Clearance of toxicants that reach the respiratory system is vitally important to proper respiratory function. Breakdown of these clearance mechanisms can have adverse effects on one's health.

STRUCTURE OF RESPIRATORY TRACT

The respiratory tract consists of three major areas: the nasopharyngeal, the tracheobronchial, and the pulmonary (see Figure 1.3).

The nasopharyngeal area consists of the nose, pharynx, and larynx. The nose consists of an external portion—which is mainly bone, cartilage, and tissue—and an internal portion. This internal portion is composed of two wedge-shaped cavities separated by a septum. The internal portion is covered by a thick mucous membrane that warms and moistens inhaled air. The nose filters in two ways: (1) the hairs that can be seen in the nose filter out the coarsest foreign materials, and (2) air currents pass over the moist mucous membranes in curved pathways and deposit fine particles against the wall. The trapped particles are subsequently carried to the pharynx and swallowed.

The pharynx is a musculomembranous tube, five inches in length, extending from the base of the skull to the esophagus. The pharynx is divided into three parts: nasal, oral, and laryngeal. The pharynx serves as a passage for two systems—the respiratory and digestive systems. The larynx, or voice box, connects the pharynx with the trachea. Its opening is at the base of the tongue. The larynx consists of nine cartilages united by muscles and ligaments. The chief function of

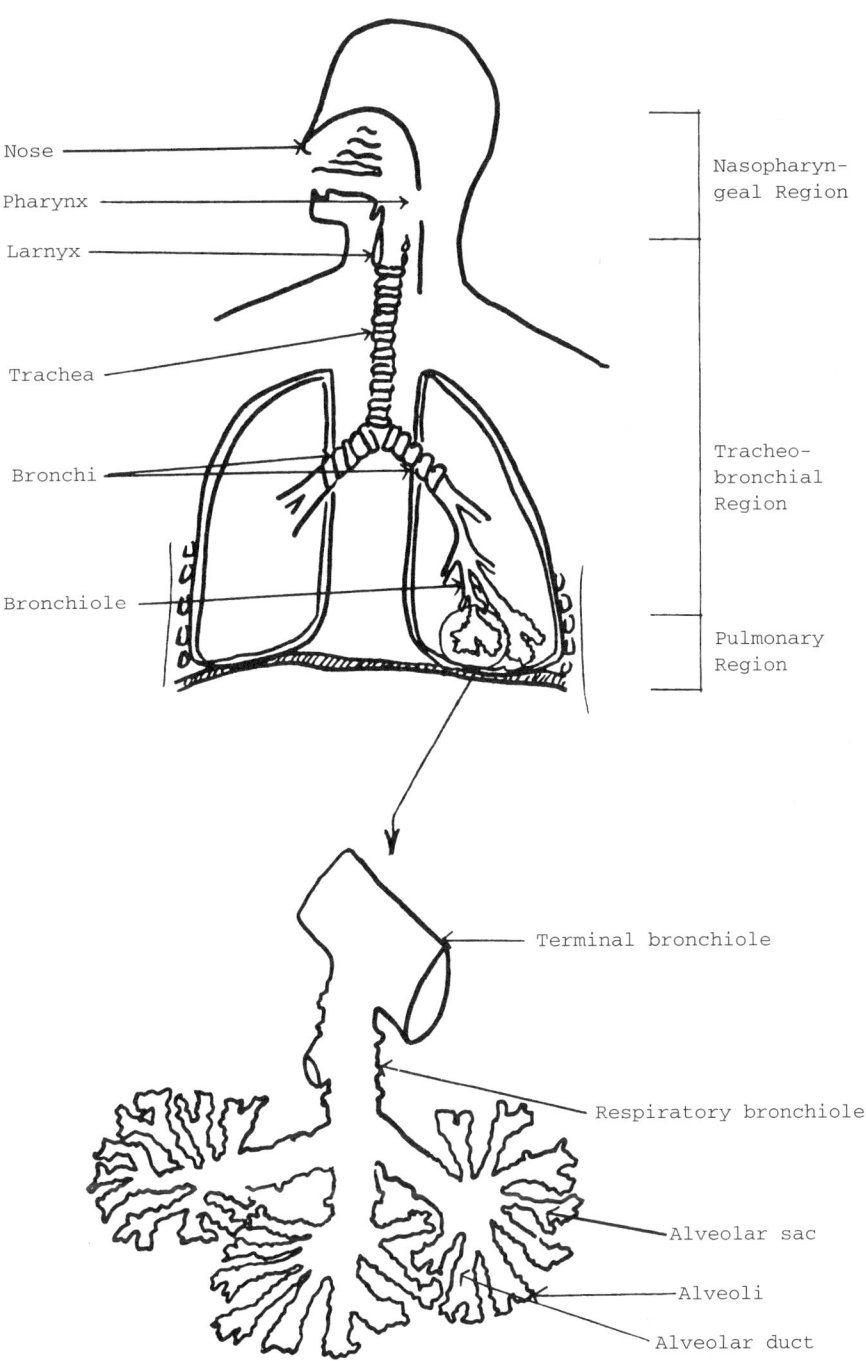

Figure 1.3   Schematic representation of the respiratory tract.

the larynx is to vocalize. The pitch of the sound is determined by the shape and tension of the vocal cords. The voice is refined by the nose, mouth, and pharynx, as well as the sinuses, which act as sounding boards and resonating chambers.

Respiration starts when air is inhaled through the nose. The main function of the nasopharynx is to warm and moisten air and to filter and remove the largest particles (larger than 10 µm) from the air inhaled. The surfaces of the nose, sinus, and upper bronchi are covered with mucous membranes. These membranes secrete a fluid called *mucus* that is produced continuously and drains slowly into the throat. The mucus provides heat and humidity to the air chambers and thus warms incoming air. In addition to mucus, the membrane is coated with cilia (hairlike filaments), which wave back and forth. Millions of cilia line the naso-pharyngeal area and aid in cleaning incoming air and removing deposited materials.

The tracheobronchial area consists of the trachea, bronchi, and bronchiole. The trachea, or windpipe, is a cylindrical tube about four to five inches in length, consisting of cartilage separated by fibrous and muscular tissue. The trachea functions as a simple passageway for air passing to the lungs.

The bronchi consist of two separate passages that split off at the base of the trachea. The right bronchus differs from the left in that it is shorter and wider and takes a more vertical course. Because of this, foreign material tends to follow the right bronchus more often than the left. The bronchi become narrower as they approach the lungs. With this narrowing, the cartilage of the bronchi tends to be reduced and disappears at the bronchiole. The epithelial lining of the trachea loses its cilia and decreases in size.

The respiratory bronchiole are tubular with lengths of approximately one-half inch. Many bronchiole branch off the bronchi. The walls are void of cartilage and contain only limited cilia. The tracheobronchial areas serve as conducting airways between the nasopharynx and the alveoli, where the gas exchange takes place. As in the nasopharynx, the tracheobronchial passageways are lined with cilia and coated with a thin layer of mucus. Medium-sized particles (1–10 µm) are filtered out in this region of the respiratory tract. The surface of these airways serves as a mucociliary escalator, moving particles from the deep lung to the oral cavities so they can be swallowed or excreted.

The pulmonary section consists of the alveolar duct, the alveolar sac, and the alveolus. Gas exchange takes place here between alveoli and blood capillaries. Particles less than 1 µm in size are deposited in this area. Figure 1.4 illustrates volumes in respiration and represents the normal respiratory pattern at rest. Changes in volume and flow are recorded using a spirometer, which determines impairment to the respiratory system. The four primary lung volumes are:

1. *Tidal volume (TV)*. The volume of gas inspired or expired during each respiratory cycle. Normally only a small volume of the lung is ventilated. The tidal volume is normally 500 ml.
2. *Inspiratory reserve volume (IRV)*. The maximum volume that can be forcibly inspired after a normal inspiration. Typically this is 3300 ml for men.

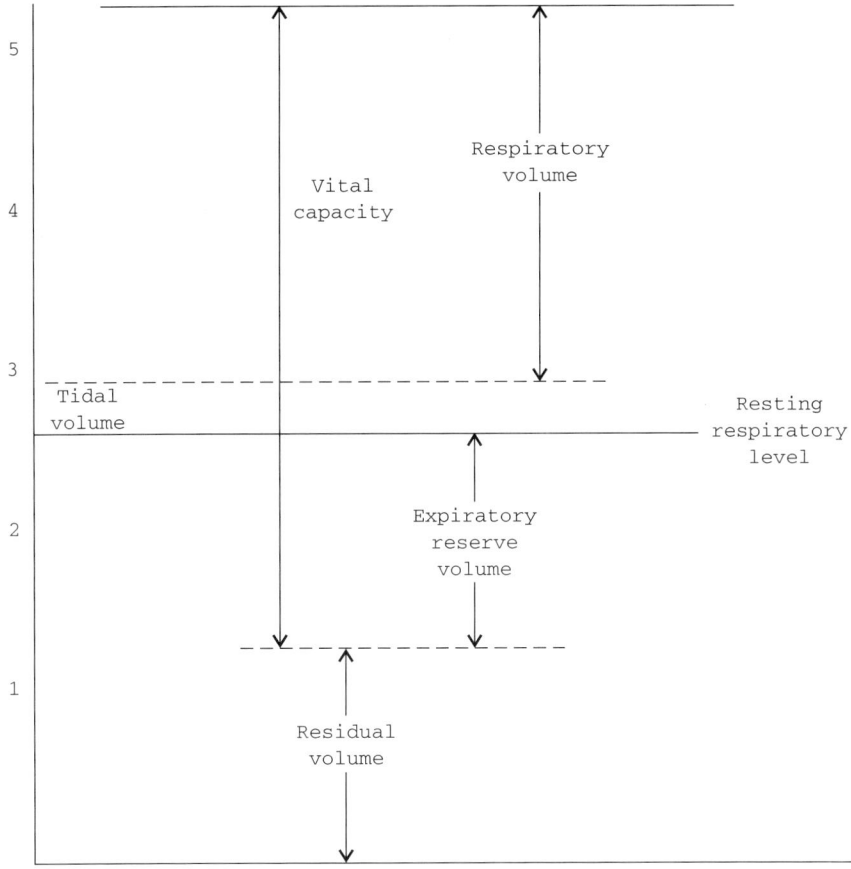

Figure 1.4   Lung volumes.

3.   *Expiratory reserve volume (ERV)*. The maximum amount of air that can be forcibly expired after normal expiration. For men this is normally 1000 ml.
4.   *Residual volume (RV)*. The amount of air remaining in the lungs after a maximum expiratory effort. Typically 1200 ml.

Other volumes that make up several of the primary ones are:

1.   *Total lung capacity (TLC)*. The sum of all four volumes listed previously.
2.   *Inspiratory capacity (IC)*. The maximum volume of air that can be inspired at the end of a normal tidal expiration (3800 ml is average for men) (IC = IRV + TV).
3.   *Vital capacity (VC)*. The maximum amount of air that can be exhaled from the lungs after a maximum inspiration (4800 ml is typical for men) (VC = IRV + TV + ERV).

4. *Functional residual capacity (FRC)*. The normal volume at the end of passive exhalation (i.e., the gas volume that normally remains in the lung and functions as the residual capacity).

The flow of air from the lungs can be used to diagnose changes in the lung. If the subject inhales and exhales maximally, the forced vital capacity (FVC) and forced expiratory volume at one second (FEV_1) can be recorded. The FVC is the maximal volume of air that can be exhaled forcefully after a maximal inspiration. The FEV_1 is the volume of air that can be forcibly expelled during the first second of expiration. Inhalation of toxicants may change the ability of the lungs to move air in and out because of obstruction or restriction of air passages. FVC and FEV_1 measurements can be used to detect these obstructions or restrictions in the lungs.

RESPIRATION RATE

Factors that can affect the rate of respiration include the following:

● Partial pressure of carbon dioxide
● Partial pressure of oxygen
● pH of arterial blood

Increased partial pressure of carbon dioxide in arterial blood increases respiration, whereas decreased partial pressure of carbon dioxide decreases the respiration rate. On the other hand, decreased partial pressure of oxygen in arterial blood increases respiration. A low arterial pH increases the respiration rate. Pain, sudden cold, and other sensory stimuli can elicit respiratory changes. Age is another factor influencing respiration rate. At birth, the rate is rapid (from 40–70 times/minute). This rate decreases with age, so that at 1 year it is 35–40 times/minute; at 10 years, 20 times/minute; and at 25 years, 16–18 times/minute. With old age, the rate can increase again to more than 20 times/minute. Body temperature affects the rate of respiration; increased body heat increases respiration.

ACTION OF TOXICANTS IN RESPIRATORY SYSTEM

Once the toxic material has been inhaled, its site of action and effect depend on several factors, including chemical composition, chemical state, solubility, and reactivity. Inhaled contaminants that adversely affect the lungs fall into two major categories:

1. *Gases and vapors*. Invisible contaminants mixed in the air. *Gases* are substances that become airborne at room temperature. Gases are often produced by chemical processes and high-heat operations. They drift quickly and undetected from their source. Nitrogen, chlorine, carbon monoxide, carbon dioxide,

and sulfur dioxide are examples. *Vapors* are formed when liquids or solids evaporate, typically occurring with solvents, paints, or refining activities.

2. *Particulate matter*. Inhaled particulates can take many forms, commonly called *aerosols*. An aerosol is a suspension of solid particles or liquid droplets in a gaseous medium. There are at least three forms of particulate materials that we must be concerned with:

- *Fumes*. An aerosol created when solid material is vaporized at high temperatures and then cooled. As it cools, it condenses into extremely small particles, generally less than 1 micron in diameter. Fumes can result from operations such as welding, cutting, smelting, or casting molten metals.
- *Dusts*. An aerosol consisting of mechanically produced solid particles derived from breaking up larger particles. Dusts generally have a larger particle size when compared to fumes. Operations such as sanding, grinding, crushing, drilling, machining, or sand blasting are the worst dust producers. Dust particles are often found in the harmful size range of 0.5–10 microns.
- *Mists*. An aerosol formed by liquids, which are atomized and/or condensed. Mists can be created by operations such as spraying, plating, or boiling, and by mixing or cleaning jobs. Particles are usually found in the size range of 5–100 microns.

Gases and Vapors

Toxic gases and vapors may be inhaled and distributed throughout the entire respiratory system. While the exchange of oxygen and carbon dioxide can take place only deep in the lungs at the alveoli, intake of toxic gases and vapors can take place anywhere in the respiratory tract. The diffusion of these toxic materials is the driving force behind the interaction of these gases and vapors within different regions of the respiratory tract. Usually there is a very small concentration of the toxicant in the tissues, while there are higher concentrations in the inspired air. Diffusion is the phenomenon of a material moving from a high concentration to a low concentration. Thus, the concentration differential allows toxic gases and vapors to move from air to the tissues of the respiratory tract. The solubility of these toxic materials in water greatly influences the relative toxicity and the site of reaction in the respiratory tract. Very soluble gases, such as ammonia or sulfur dioxide, are absorbed in the nasopharyngeal area, whereas other gases, such as nitrogen dioxide and organic solvent vapors, are less soluble in water. The latter do not readily dissolve in the mucous membrane and are able to reach lower into the airways and can even reach the alveolar area, resulting in their diffusion into the bloodstream. In this latter category of toxicant, the direct effect may not be on the lungs; rather, their diffusion into the bloodstream may mean that other organs will be affected. Those toxicants that are very water-soluble react with the mucous layer of the respiratory tract. Because the mucous layer is always being renewed and removed by ingestion or carried to the mouth by the cilia, dissolution in this layer detoxifies the gas or vapor.

The lining of the upper portion of the respiratory tract consists of mainly goblet and ciliated cells. Goblet cells are responsible for secreting mucus. If the toxic gas or vapor is capable of interfering with the goblet cells, then the rate of mucous secretion is affected. Once the toxicant has penetrated the mucous lining of the upper airways (highly water-soluble materials), then the goblet and ciliated cells can be attacked. It appears that ciliated cells are more prone to the effects of toxic gases and vapors. Cilia may be lost, and their cells may even die. This results in a lack of cilia in the area, and the process that involves removing foreign materials can be greatly affected. If the toxic gas or vapor is not very soluble in water, it will most likely reach the lower airways (alveoli) of the lungs. Once this occurs, intake of the gas or vapor is the same as that for oxygen (i.e., diffusion determines the fate of the toxicant). The alveoli are separated from the bloodstream by a very thin lining, and there are usually lower concentrations of the toxicant in the blood than in the alveoli; thus, via diffusion, gases and vapors are capable of moving into the bloodstream. Even though toxic gases or vapors reach the lower airways, it does not mean that they will pass into the blood and not have toxic effects on the lower portion of the respiratory tract. Certain regions of the lower airway are more affected by gases and vapors that act directly on the lung. For example, ozone affects the region between the bronchiole and the alveolus more so than other areas of the body.

The respiratory system may also be exposed to toxic gases and vapors through its function in the excretory process. Vapors from organic compounds, such as benzene and toluene, are excreted through the lungs. If the metabolism of these vapors to more water-soluble metabolites is slow, then high levels of vapors may be exhaled. Toxic materials absorbed by other routes—or toxic metabolites from other areas in the body—may be excreted through the respiratory system. The toxicants may then accumulate in a high concentration in the lungs, resulting in pulmonary damage.

Particulate Material

The site of deposition in the respiratory tract plays a major role in the toxic effect of the inhaled aerosol. Several factors determine the site of deposition of the aerosol, namely size, density, shape, and tendency to aggregate. Particle size plays the major role in determining where the aerosol will settle out in the respiratory tract. Usually, the inhaled particles are not uniform in size and shape. For simplicity, we consider all particles to be spheres of unit density, and the size distribution of the aerosol approximates a lognormal distribution. We usually refer to the particle size of the aerosol as the median or geometric mean particle size. Also, reference is usually made to the aerodynamic diameter of a particle that takes into account the density of the particle and the aerodynamic drag. The aerodynamic diameter represents the diameter of a unit-density sphere with the same terminal settling velocity as the particle, whatever its size, shape, and density. Aerosols can range in size from 0.01–100 μm. Figure 1.5 illustrates the type of aerosols and their typical particle size.

The American Conference of Governmental Industrial Hygienists (ACGIH) TLVs booklet[1] classifies airborne particulate into three categories (see Tables 1.1 to 1.3):

- *Inhalable particulate mass*. Materials that are hazardous when deposited anywhere in the respiratory tract.
- *Thoracic particulate mass*: Materials that are hazardous when deposited within the lung airways and the gas-exchange region.
- *Respirable particulate mass*: Materials that are hazardous when deposited in the gas-exchange region.

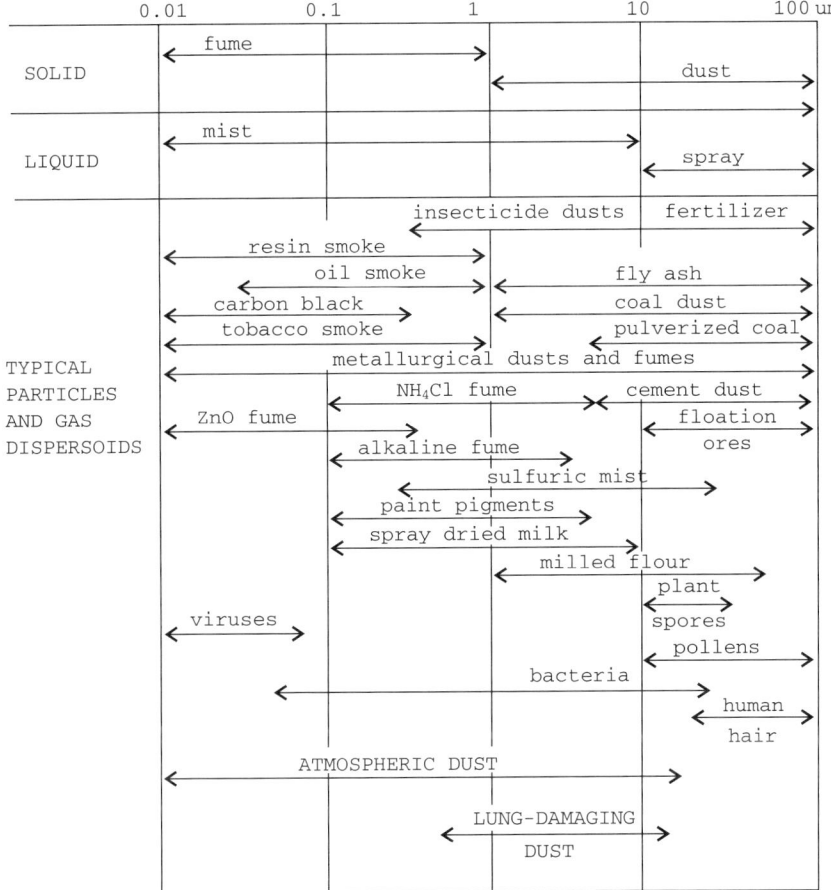

Figure 1.5 Sizes of airborne particles.

Table 1.1   Inhalable particulate mass

PARTICLE AERODYNAMIC DIAMETER (μM)	INHALABLE PARTICULATE MASS (%)
0	100
1	97
2	94
5	87
10	77
20	65
30	58
40	54.5
50	52.5
100	50

Table 1.2   Thoracic particulate mass

PARTICLE AERODYNAMIC DIAMETER (μM)	THORACIC PARTICULATE MASS (%)
0	100
2	94
4	89
6	80.5
8	67
10	50
12	35
14	23
16	15
18	9.5
20	6
25	2

Table 1.3   Respirable particulate mass

PARTICLE AERODYNAMIC DIAMETER (μM)	RESPIRABLE PARTICULATE MASS (%)
0	100
1	97
2	91
3	74
4	50
5	30
6	17
7	9
8	5
10	1

Four mechanisms influence the deposition of particulate material in the respiratory tract: interception, impaction, sedimentation, and diffusion.

- *Interception* occurs when the movement of the particle in the airways brings it into close proximity with the surface of the respiratory tract. By being in close proximity with the surface, the particle is able to come into contact with the surface and is removed from the airstream. Interception plays a major role in particle filtering in the upper nasopharyngeal region and affects particles that are larger than 30 μm.
- *Impaction* occurs when the airways make abrupt changes in direction, as in the lower nasopharyngeal region. As the airways change direction, by momentum, the particle is forced to move in one direction and thus runs into the surface of the passageway. Particles in the 5–30 μm size range are strongly influenced by impaction and filtered out in the lower nasopharyngeal region. Airflow in the nasopharyngeal region is very high (more than 180 cm/sec). This high airflow—and the tortuous nature of the air passages in this region—causes the airflow to change direction often. This provides the ideal condition for impaction to take place.
- *Sedimentation* brings about the deposition of particles of the 1–5 μm size into the smaller bronchi and the bronchioles, where the velocity of airflow is low. As the particles move down the airways, buoyancy and resistance of air act on the particles in an upward direction while the gravitational forces act in a downward direction. Because the velocity of the air decreases as it moves deeper into the lungs, the gravitational forces equal or exceed the sum of the buoyancy and air resistance. This results in the particles settling out on the surface of the bronchi and bronchioles.
- *Diffusion* is important in the aveolar region. Because the velocity of airflow in this region is very low, the particles begin to collide with gas molecules. Random motion is important to the particles in this process, which is referred to as *Brownian motion*. Brownian motion increases with decreasing particle size, and its random motion causes the particles to come into contact with the aveoli surface and become deposited there. Particles of less then 1 μm size distribution are affected by diffusion.

Breathing patterns influence particle deposition. During quiet breathing, in which the tidal volume is only two or three times the volume of the dead space, a large proportion of the inhaled particles may be exhaled. During exercise or heavy work, larger volumes of air are inhaled at higher velocities; thus, the deposition mechanisms increase. Holding one's breath can also increase deposition by sedimentation and diffusion. Any constriction in the airways caused by irritant materials, sickness, diseases, and so on, will increase the amount of particle deposition by impaction.

The respiratory tract is capable of removing deposited particulate material from the site of deposition. This clearance mechanism varies depending on the site of deposition. In the nasopharyngeal and tracheobronchial regions, the air passages are coated with cilia, which move back and forth at all times. The particles trapped in the mucous coating of those regions are transported by a mucociliary escalator up to the mouth and are ingested.

Particulate material that is deposited in the alveolar region is subjected to three main avenues of clearance:

1. Particles being phagocytized and cleared via the mucociliary escalator up the tracheobronchial tree.
2. Particles being phagocytized and removed via lymphatic drainage.
3. Partial or complete dissolving of the particle and removal via the bloodstream or lymphatic drainage.

Within the alveolar region, there are many macrophages, which engulf all particles within several hours of inhalation. The phagocytized particle is then capable of being cleared from the system via one of the three mechanisms listed previously.

In order for a toxic effect to occur in the human body, the toxicant must reach the appropriate site of action in a high enough concentration and stay there long enough to exert its toxic effect. Thus, changes in dose, route, and duration of exposure can influence the toxic effect. A detailed summary of toxicology is not discussed in this book, but readers should be aware of the variables that may influence the toxic effect of foreign material in the human body.

Factors that influence the toxicity of a material can be related to the toxic agent, the exposure, and the subject. These factors can be further divided. Table 1.4 summarizes the variables that can affect the toxicity of materials that enter the body.

In summary, the respiratory system is vulnerable to airborne contaminants. Depending on the nature of the inhaled material, the site of deposition, and so on, certain toxic effects may be felt in the respiratory system. Table 1.5 gives examples of site of action and pulmonary disease produced by inhaled toxicants. The respiratory tract is an effective filtering system and is capable of removing the contaminant from the body. Irritation of the air passages resulting in

Table 1.4 Summary of factors that influence the toxicity of materials entering the human body

AGENT	EXPOSURE	SUBJECT
Chemical composition	Dose	Age, sex, body weight
Physical composition	Route of entry	Health status
Solubility in biologic fluids	Duration and frequency	Nutritional status
Presence of other materials		Genetic status
		Environmental factors such as temperature, humidity, social factors, geographic influences

Table 1.5   Site of action and pulmonary disease produced by selected occupationally inhaled toxicants

TOXICANT	COMMON NAME OF DISEASE	SITE OF ACTION	ACUTE EFFECT	CHRONIC EFFECT
Aluminum	Aluminosis	Upper airways, alveolar interstitium	Cough, shortness of breath	Interstitial fibrosis
Aluminum abrasives	Shaver's disease, corundum smelter's lung, bauxite lung	Alveoli	Alveolar edema	Fibrotic thickening of alveolar walls, interstitial fibrosis, and emphysema
Ammonia		Upper airway	Immediate upper and lower respiratory tract irritation, edema	Chronic bronchitis
Arsenic		Upper airways	Bronchitis	Lung cancer, bronchitis, laryngitis
Asbestos	Asbestosis	Parenchyma		Pulmonary fibrosis, pleural calcification, lung cancer, pleural mesothelioma
Beryllium	Berylliosis	Alveoli	Sever pulmonary edema, pneumonia	Pulmonary fibrosis, progressive dyspnoea, interstitial granulomatosis, cor pulmonale
Boron		Alveolus	Edema and hemorrhage	
Cadmium oxide		Alveolus	Cough, pneumonia	Emphysema, cor pulmonale
Cake oven emissions		Upper airways		Tracheobronchial cancers
Carbides of tungsten, titanium, tantalium	Hard metal disease	Upper airway and lower airway	Hyperplasia and metaplasia of bronchial epithlium	Fibrosis, cor pulmonale, peribronchial and perivascular fibrosis

Table 1.5 (*continued*)

TOXICANT	COMMON NAME OF DISEASE	SITE OF ACTION	ACUTE EFFECT	CHRONIC EFFECT
Chlorine		Upper airways	Cough, hemoptysis, dyspnoea, tracheobronchitis, bronchopneumonia	
Chromium (VI)		Nasopharynx, upper airways	Nasal irritation, bronchitis	Lung tumors and cancers
Coal dust	Pneumoconiosis	Lung parenchyma, lymph nodes, hilus		Pulmonary fibrosis
Cotton dust	Byssionsis	Upper airways	Tightness in chest, wheezing, dyspnea	Reduced pulmonary function, chronic bronchitis
Hydrogen fluoride		Upper airways	Respiratory irritation, hemorrhagic pulmonary edema	
Iron oxides	Siderotic lung disease: Silver finisher's lung, hematite miner's lung, arc welder's lung	Silver finisher's: pulmonary vessels and alveolar walls; hematite miner's: upper lobes, bronchi and alveoli; arc welder's: bronchi		Silver finisher's: subpleural and perivascular aggregations of macrophages; hematite miner's: diffuse fibrosislike pneumonconiosis; arc welder's: bronchitis Pulmonary fibrosis
Kaolin	Kaolinosis	Lung parenchyma, lymph nodes, hilus		Pulmonary fibrosis
Manganese	Manganese pneumonia	Lower airways and alveoli	Acute pneumonia, often fatal	Recurrent pneumonia

TOXICANT	COMMON NAME OF DISEASE	SITE OF ACTION	ACUTE EFFECT	CHRONIC EFFECT
Nickel		Parenchyma (NiCO), nasal mucosa (Ni_2S_3), bronchi (NiO)	Pulmonary edema, delayed by two days (NiCO)	Squamous cell carcinoma of nasal cavity and lung
Osmium tetraoxide		Upper airways	Bronchitis, bronchopneumonia	
Oxides of nitrogen		Terminal respiratory bronchi and alveoli	Pulmonary congestion and edema	Emphysema
Ozone		Terminal respiratory bronchi and alveoli	Pulmonary edema	Emphysema
Phosgene		Alveoli	Edema	Bronchitis
Perchloroethylene			Pulmonary Edema	
Silica	Silicosis, pneumoconiosis	Lung parenchyma, lymph nodes, hilus		
Sulfur dioxide		Upper airways	Bronchoconstriction, cough, tightness in chest	
Talc	Talcosis	Lung parenchyma, lymph nodes		Pulmonary fibrosis
Tin	Stanosis	Bronchioles and pleura		Widespread mottling of x-ray without clinical signs
Toluene		Upper airways	Acute bronchitis, bronchospasm, pulmonary edema	
Vabadium		Upper and lower airways	Upper airway irritation and mucus production	Chronic bronchitis
Xylene		Lower airways	Pulmonary edema	

Table 1.6   Oxygen deficiency and effects

PERCENT	OXYGEN PHYSIOLOGICAL EFFECT
19.5-16	No visible effect
16-12	Increased breathing rate; accelerated heartbeat; impaired attention, thinking, and coordination
12-10	Faulty judgment and poor muscular coordination; muscular exertion causing rapid fatigue; intermittent respiration
10-6	Nausea and/or vomiting; inability to perform vigorous movement or loss of the ability to move; unconsciousness, followed by death
Below 6	Difficulty in breathing; convulsive movements; death in minutes

(1) constriction of the airways; (2) damage to the cells lining the airways; (3) production of fibrosis; (4) constriction of the airways caused by allergic reactions; and (5) production of lung tumors are the five general toxic responses of the respiratory system to airborne contaminants.

Oxygen Deficiency

Oxygen deficiency—a lack of oxygen in the air—can be caused by chemical reactions, fire, or displacement by other gases. In confined spaces, where ventilation is limited or nonexistent, aerobic bacterial growth and oxidation of rusting metals can also cause an oxygen-deficient atmosphere. Oxygen constitutes only a small percentage, about 21%, of the air we breathe. Yet, when levels of oxygen fall below 19.5% (minimal acceptable level), life-threatening health problems begin to occur quickly. Oxygen deficiency is a serious situation that can cause loss of consciousness or death in minutes. The impact of oxygen deficiency can be gradual or sudden. Typically, decreasing levels of atmospheric oxygen cause the physiologic symptoms shown in Table 1.6.

SOURCES OF RESPIRATORY HAZARDS

The wide variety of processes used in industry create numerous potential health problems. In this section, examples of chemical substances associated with industrial processes are listed in Table 1.7. The list is by no means complete, for either processes or chemical contaminants, and is intended only as a guide.

Table 1.7   Air contaminants from industrial processes

PROCESS TYPE	CONTAMINANT TYPE	EXAMPLES OF CHEMICAL CONTAMINANTS
Abrasives (manufacture)	Dust	aluminum oxide, silicon carbide silica, emery, corundum
	Gas/vapors	carbon monoxide, solvent vapors from adhesives, vaporized resins
Adhesives (manufacture and use)	Vapors	solvent vapors, resins
Asphalt paving	Dust	silica, silicates, carbonates
	Vapors	polycrylic, aromatic hydro-carbons, aromatic and aliphatic hydrocarbon solvents
Automotive (repair and maintenance)	Dust/ fibers	Asbestos, metal and resin dust from grinding
	Fumes	metal oxides (welding)
	Gas/vapors	petroleum solvents, gasoline, carbon monoxide, hydrogen, nitrogen oxides, styrene, acetone, isocyanates, organic peroxides
Bakeries	Dust	flour, other vegetable dust, yeast, molds
Battery manufacture	Dust/fume	lead, calcium
	Gas	hydrogen, formaldehyde, vinyl chloride
	Mists	sulfuric acid, hydrochloric acid, alkali mists

Table 1.7   *(continued)*

PROCESS TYPE	CONTAMINANT TYPE	EXAMPLES OF CHEMICAL CONTAMINANTS
Beverage and soft drink manufacture	Gas	ammonia, carbon dioxide
	Mists	caustic mists
Blasting, abrasive	Dust	silica, silicates, carbonates, lead, cadmium, zinc
Boiler making	Dust	silicates, fluorides, carbonates
	Fume	welding, fumes, metal fumes
Brewing	Gas	refrigerant gases, carbon dioxide
	Mist	caustic mists
Brick and tile manufacture	Dust	silica, silicates, fluorides, carbonates
	Gas (kilns)	carbon monoxide
Business machines (photocopying and duplicating)	Gas	ammonia, ozone
	Vapors	methyl alcohol, chlorinated hydrocarbons and petroleum solvents
Can manufacture	Fumes	metal fumes
	Vapors	solvent vapors
Cement manu-facture	Dust	silica, silicates, fluorides, carbonates, chromates

Table 1.7   *(continued)*

PROCESS TYPE	CONTAMINANT TYPE	EXAMPLES OF CHEMICAL CONTAMINANTS
Cement products industry	Dust	cement
	Fumes	welding fumes
	Gas/ vapors	gasoline, acetone, lacquer thinner, kerosene, fuel oil, Stoddard solvent
Charcoal production	Gas	carbon monoxide, polycyclic aromatic hydrocarbons
Chemical manufacture: Acid plants	Gas	sulfur dioxide, nitrogen oxides
	Mists	acid mists
Benzoic acid	Dust	benzoic acid
Chlor-alkali plant	Gas/vapor	chlorine, mercury
	Mist	sodium hydroxide
	Dust	fluoride, phosphate, silicates, carbonates, distomacious earth
Fertilizer	Gas	ammonia, hydrogen fluoride
	Mist	phosphoric acid
Solvents	Vapors	solvent vapors—alcohols, ketones, esters, aliphatic hydrocarbons, aromatic hydrocarbons, chlorinated hydrocarbons
Clay	Dust	mica, silicates, iron oxide, silica (quartz)

Table 1.7   *(continued)*

PROCESS TYPE	CONTAMINANT TYPE	EXAMPLES OF CHEMICAL CONTAMINANTS
Coal handling	Dust	coal dust, silica
	Gas	sulfur dioxide, carbon monoxide
Coke handling	Dust	coke dust
Coking	Gas	carbon monoxide, ammonia, hydrogen sulfide, sulfur dioxide, phenols, cyanides, naphthalene and other poly-cyclic aromatic hydrocarbons, benzene, pyridine, carbon disulfide
Dairy processing industry	Mist	alkali mists
Dental industry	Vapor	mercury
Detergent manufacture and use	Dusts	proteolytic enzymes, sodium perborate, phosphates
	Mists	alkali mists
Distilleries	Vapor	alcohol (ethyl alcohol)
Drilling (rock)	Dust	silica, silicates, carbonates, fluorides
Dry cleaning	Vapor	perchlorethylene, trichlorethylene, petroleum solvents
Electrical components manufacturing industry	Fumes	metal fumes (silver, lead, cadmium, tin)
	Vapors	solvent vapors, freon gases

Table 1.7   *(continued)*

PROCESS TYPE	CONTAMINANT TYPE	EXAMPLES OF CHEMICAL CONTAMINANTS
Electropolating and galvanizing acid:	Gas	hydrogen
	Mist	chromic acid, sulfuric acid, cyanide, sulfamate, hydrochloric acid
Alkaline	Gas	ammonia
	Mist	sodium stannate (tin salt)
Cyanide	Gas	ammonia, hydrogen cyanide
	Mist	cyanide, alkali
Electrodeless plating	Gas	formaldehyde, ammonia
Electro-polishing	Gas	hydrogen fluoride, hydrogen chloride
	Mist	sulfuric acid, hydrofluoric acid, phosphoric acid, hydrochloric acid, chromic acid, perchloric acid
Fluoroborate	Mist	fluoroborate mist
Galvanizing	Dust	metal oxide
	Fumes	lead, zinc
	Gas	ammonia, hydrogen chloride
	Mist	alkali, hydrochloric acid, sulfuric acid

Table 1.7   *(continued)*

PROCESS TYPE	CONTAMINANT TYPE	EXAMPLES OF CHEMICAL CONTAMINANTS
Explosives	Gas	oxides of nitrogen, carbon monoxide, sulfur dioxide
Fiberglassing (also see plastics)	Dust	asbestos, wood dust, glass fibers, resins, glycols, peroxides
	Vapors	acetone, styrene, amines, alcohols, phthalates, methyl ethyl ketone, toluene, phenol, isocyanates
Foundries, furnaces, and forges: Basic oxygen furnace material handling	Dust	iron oxide, graphite, limestone, ore, mill scale, flouspar
Forging	Gas	sulfur dioxide, carbon monoxide, carbon dioxide
	Vapor	acrolein
Foundry operations	Dust	silica, silicates, carbonates, fluoride, cyanides
	Fumes	metal oxides
	Gas	ammonia, carbon monoxide, sulfur dioxide phosgene, chlorine, fluorine, nitrogen oxides
	Vapors	acrolein, aldehydes, phenols, isocyanates, polycyclic aromatic hydrocarbons

Table 1.7   *(continued)*

PROCESS TYPE	CONTAMINANT TYPE	EXAMPLES OF CHEMICAL CONTAMINANTS
Furnace operations (all types)	Dust/fumes	iron oxide, other metal oxides, fluxing agents
	Gas	sulfur dioxide, carbon monoxide, other combustion materials
Leaded steel making	Dust/fumes	lead oxide, iron oxide
Sintering	Dust	iron oxide, silica, fluorides, carbonates, metal oxides
	Gas	sulfur dioxide, carbon monoxide
Tandem mills	Mist	oil mists
Frozen food industry	Gas	ammonia, methyl chloride, freons
Gases, compressed	Gas	asphyxiating, corrosive or toxic gases, flammable or explosive gases
Glass industry: Etching	Gas	hydrogen fluoride
Fiberglassing (see fiberglassing) Manufacture	Dust/fumes	silica, lead, soda ash, potash, vanadium, arsenic
	Fibers	asbestos
	Gas	sulfur dioxide, hydrogen fluoride
	Mist	oil mists

Table 1.7   *(continued)*

PROCESS TYPE	CONTAMINANT TYPE	EXAMPLES OF CHEMICAL CONTAMINANTS
Hair and bristle processing industry	Dust	fiber dust, spores of animal diseases
	Gas/Vapors	sulfur dioxide, hypochlorite vapor
Hospitals	Gas	formaldehyde, anesthetic gases, ethylene oxide
Insulation manufacturing	Dust	mineral dust, cellulose dust, silica
	Fibers	asbestos, glass
	Vapors	isocyanates
Joineries, cabinet making, and furniture manufacture	Dust	wood dust
	Vapors	solvents, glues, paints
Metal cleaning and surface treatment operations abrasive cleaners	Dust	silica, insoluble silicates, calcium carbonate, pumice, sodium carbonate, sodium silicate, di- and trisodium phosphate
Acid cleaners: Dipping	Gas	oxides of nitrogen, hydrogen, hydrogen chloride
	Mist	nitric acid, sulfuric acid, chromic acid, hydrochloric acid
Pickling	Gas	oxides of nitrogen, hydrogen fluoride, hydrogen chloride, hydrogen cyanide, hydrogen, arsine

Table 1.7   *(continued)*

PROCESS TYPE	CONTAMINANT TYPE	EXAMPLES OF CHEMICAL CONTAMINANTS
Alkaline cleaners:	Mist	alkali mists
Case hardening	Gas	carbon monoxide, oxygen deficiency, cyanides
Etching	Gas	hydrogen fluoride
	Mist	alkali mists
Degreasing	Vapors	trichlorethylene, perchlorethylene, petroleum and chlorinated hydrocarbon solvents
Strike solutions	Gas	hydrogen chloride
	Mist	cyanide, chloride
Stripping operations	Gas	hydrogen chloride, oxides of nitrogen
	Mist	hydrochloric acid, chromic acid, acetic acid, nitric acid, hydrofluoric acid, cyanide, alkali, mists
Metal spraying	Dust/fumes	metals and oxides of metals (nickel, chromium, cobalt)
Paint manufacture	Fumes	lead oxide, mercuric oxide resins
	Vapors	solvents, isocyanates, polyurethanes, insecticides

Table 1.7   *(continued)*

PROCESS TYPE	CONTAMINANT TYPE	EXAMPLES OF CHEMICAL CONTAMINANTS
Paperboard/ container industry	Fumes	welding fumes
	Gas	formaldehyde, carbon monoxide
	Vapors	gasoline, acetone, lacquer thinner, kerosene, fuel oil, stoddard solvent
Pest control (fungicides, herbicides, pesticides, rodenticides)	Dust/vapors	prganophosphorus compounds, halogenated hydrocarbons, lead arsenate, carbonates, thiocarbonates, dinitrocresol, thallium and its compounds, coumarin, indane and derivatives, chloropicrin, mercury, creosote, dinitrophenol, solvents
Petroleum refineries	Gas	hydrogen sulfide, mercaptans, liquefied petroleum gases
	Vapors	solvent vapors
Photographic industry	Dust	organic dyes
	Vapors	aminophenols, hydroquinone, acetic acid
Plastics and resins	Gas	thermal decomposition products, (carbon monoxide, carbon dioxide, oxides of nitrogen), blowing agents
	Vapors	isocyanates, monomers (e.g., styrene, vinyl chloride)

Table 1.7   *(continued)*

PROCESS TYPE	CONTAMINANT TYPE	EXAMPLES OF CHEMICAL CONTAMINANTS
Plumbing, heating, and air conditioning contractors	Fibers	asbestos, glass
	Fumes	welding fumes
	Gas	carbon monoxide, refrigeration gases
	Mist	acid mists, caustic mists
	Vapors	solvents
Pottery and porcelain industry	Dust	clay, silica, silicates
	Fumes	lead
Power stations, thermal	Dust	vanadium oxide, nickel
	Gas	sulfur dioxide
	Mist	oil
Printing	Dust/fumes	lead, chromium compounds, antimony, nickel salts
	Mist	chromic acid, alkalis
	Vapors	solvents (turpentine, benzene, toluene, xylene, alcohols)
Pulp and paper industry bleaching	Gas	chlorine, chloride dioxide, sulfur dioxide

Table 1.7   *(continued)*

PROCESS TYPE	CONTAMINANT TYPE	EXAMPLES OF CHEMICAL CONTAMINANTS
Chlorine dioxide generation	Gas	sulfur dioxide, chlorine dioxide, chlorine
	Mist	sulfuric acid mist
Digesting	Gas	methyl mercaptan, dimethyl sulfide, dimethyl disulfide, hydrogen sulfide
Lime kiln	Dust	calcium oxide, calcium carbonate, sodium oxide
	Gas	carbon dioxide
	Mist	alkali mist
Recovery furnaces	Dust	sodium sulfate
	Gas	Carbon dioxide, sulfur dioxide, hydrogen sulfide, mercaptans
Refrigeration plants	Gas	ammonia, hydrogen chloride, ethane, chlorine, methyl chloride, phosgene, sulfur dioxide, ethyl chloride, propane, butane, ethylene, freons
Rock crushing and drilling	Dust	silica, silicates, carbonates, fluorides
	Gas	internal combustion engine exhaust gases (oxides of nitrogen, sulfur dioxide, carbon monoxide, aldehydes)

Table 1.7   *(continued)*

PROCESS TYPE	CONTAMINANT TYPE	EXAMPLES OF CHEMICAL CONTAMINANTS
Roofing industry	Dust/fibers	asbestos, cement dust
	Fumes	metal oxides
	Vapors	solvents, asphalt
Rubber industry: Synthetic	Dust/fumes	acetic acid, sulfuric acid
	Fibers	acrylonitrile, benzene, butadiene, chlor-butadiene, isocyanates, styrene, ethyl benzene, isoprene, dichloroethane
Vulcanizing	Dust	sulfur dioxide
	Mist	organic solvents
Sawmilling and planing filing room	Dust/fumes	cadmium, metals, metal oxide, mineral dust, welding fumes
	Fibers	asbestos
Wood rooms	Dust	wood dust
	Mist	oil mist
Scrap metal processors	Fumes	lead, cadmium, mercury, zinc, welding fumes
	Gas	fluorine
	Vapors	solvents

Table 1.7   *(continued)*

PROCESS TYPE	CONTAMINANT TYPE	EXAMPLES OF CHEMICAL CONTAMINANTS
Sewage disposal and treatment	Gas	methane, carbon dioxide, hydrogen, sulfide, mercaptans
Sewers	Gas	oxygen deficiency, methane, carbon monoxide, hydrogen sulfide, carbon dioxide, liquefied petroleum gas
	Vapors	gasoline, petroleum solvents
Shipbuilding	Dust/ Fibers	asbestos, metal oxides
	Fumes	lead, organotin and organomercurial anti-fouling paints, welding fumes
	Gas	combustion products (carbon monoxide, oxides of nitrogen)
Sign and advertising display manufacturers	Fibers	asbestos
	Fumes	welding fumes
	Vapors	methylene chloride, methyl ethyl ketone, methanol, xylene, mercury
Silos	Gas	oxygen deficiency, oxides of nitrogen

Table 1.7   *(continued)*

PROCESS TYPE	CONTAMINANT TYPE	EXAMPLES OF CHEMICAL CONTAMINANTS
Slaughtering plants (digesters and rendering)	Dust	sodium hydroxide, sodium carbonate
	Gas	chlorine, refrigerant gases
	Mist	sulfuric acid, phosphoric acid, acetic acid
	Vapors	formaldehyde, phenol- or cresol- based sanitizers
Smelting and refining	Dust/fumes	metal oxides, selenium, tellurium, cadmium, arsenic, lead, bismuty, indium, silver, gold, fluoride, metal fumes and dusts
	Gas	arsine, sulfur dioxide, carbon monoxide, hydrogen selenide, hydrogen fluoride
	Mist	sulfuric acid, fluorosilicic acid
Soldering	Fumes	metal fumes (silver, cadmium, lead, tin)
	Vapors	formaldehyde, acrolein, aldehydes
Sterilization processes	Gas	ozone, ethylene oxide, halogenated hydrocarbons

Table 1.7   *(continued)*

PROCESS TYPE	CONTAMINANT TYPE	EXAMPLES OF CHEMICAL CONTAMINANTS
Tunnelling (underground)	Dust	silica, silicates, fluorides, carbonates
	Gas	internal combustion engine exhaust gases, oxygen deficiency, oxides of nitrogen, carbon monoxide, natural gases (methane), sulfur dioxide, aldehydes, explosive byproducts
Water supply and treatment	Gas	chlorine, ammonia, hydrogen chloride, ozone
	Mist	hydrochloric acid, sodium hydroxide, calcium hydroxide
Welding and thermal cutting	Dust/fumes	metal oxides, welding fumes, silicates, carbonates, fluorspar, metal fumes, fluoride
	Gas	oxides of nitrogen, ozone, phosgene, carbon monoxide, hydrogen flouride
Wineries	Gas	carbon dioxide, refrigerant gases, sulfur dioxide
	Vapor	ethyl alcohol
Wood processing and treatment	Dust	arsenic and copper compounds
	Vapors	pentachlorophenol, tetrachlorophenol

Table 1.7   *(continued)*

PROCESS TYPE	CONTAMINANT TYPE	EXAMPLES OF CHEMICAL CONTAMINANTS
Plywood and veneer mills: Filing room	Dust/fibers	metal dust (cadmium, tin, antimony, copper, lead), asbestos
Maintenance shops	Fumes	welding fumes
	Gases	carbon monoxide, nitrogen oxides
Paint and stains	Vapors	organic solvents, pigments (compounds of titanium, iron, lead, cadmium, chromium)
Sawing and sanding	Dust	wood dust, wood preservatives
Veneer drying, glueing, and patching	Dust	resins, catalysts
	Gas	ammonia, formaldehyde
	Vapors	isocyanates, methylene chloride, urea, phenol, terpenes, alcohols, aldehydes, esters, ketones

   The terms used in the table have been defined earlier in this chapter; however, they are described again as follows for easy reference:

1. *Dust*. Finely divided solid particles dispersed in air, which have been generated by mechanical processes such as grinding, crushing, blasting, and drilling.
2. *Fume*. Solid particles dispersed in air formed by condensation of vapors of solid materials.
3. *Gas*. An aeriform that does not become liquid or solid at ordinary temperature and pressure. Gases easily diffuse into other gases and distribute throughout any container readily and uniformly.

4. *Mist.* A finely divided liquid dispersed in air, which has been generated by mechanical agitation such as splashing, pouring, or chemical reactions.
5. *Vapor.* A gaseous state of a substance that is liquid or solid at ordinary temperature and pressure.

RESPIRATORY HAZARD IDENTIFICATION AND EVALUATION

The U.S. Occupational Safety and Health Administration (OSHA) and several other safety and health regulatory bodies have recognized the need for respiratory hazard evaluation.[2] OSHA 29 CFR 1910.134 states: "The employer shall identify and evaluate the respiratory hazard(s) in the workplace; this evaluation shall include a reasonable estimate of employee exposure to respiratory hazard(s)." This necessitates periodic monitoring of the air contaminant concentration to which the respirator wearer is exposed.

Proper identification of respiratory hazards and their evaluation is needed for the following reasons:

- To determine an atmosphere's oxygen content.
- To determine the extent of exposure to the toxic substances.
- To demonstrate compliance with the legal exposure limits.
- To assess the effectiveness of engineering and administrative controls.
- To determine if the level or concentration indicates the need for additional respiratory protection.

In addition, this information is also used for correct selection of respirators for specific hazards, as shown in Figure 1.6.

Hazards are usually identified by conducting a preliminary survey. The purpose of this survey is to document the materials used or produced, how they are handled, general conditions, and the persons who may be exposed to a hazard.

The variety of materials in the workplace includes all raw materials, products, and by-products. Depending on the process, the exposure to any or all substances could be important. For example, paint thinner is a raw material in paint manufacturing. A spray painter is therefore exposed to paint thinner vapors. Undesirable reaction products are formed during certain process operations, such as the formation of zinc oxide fumes during welding and torch-cutting galvanized steel. Another example is the inadvertent breakdown of chemical compounds into more toxic substances, such as the formation of phosgene gas by welding near a vapor degreaser containing trichloroethylene solvent.

Recognizing a hazard is easier if pure chemicals are used in the process. When proprietary mixtures of chemicals are encountered, the identification of a hazard may be difficult if the manufacturer is not willing to supply information about the products. Having compiled an inventory of raw materials, products, and by-products, it would then be necessary to review the relevant toxicologic information outlined in the Material Safety Data Sheets (MSDSs).

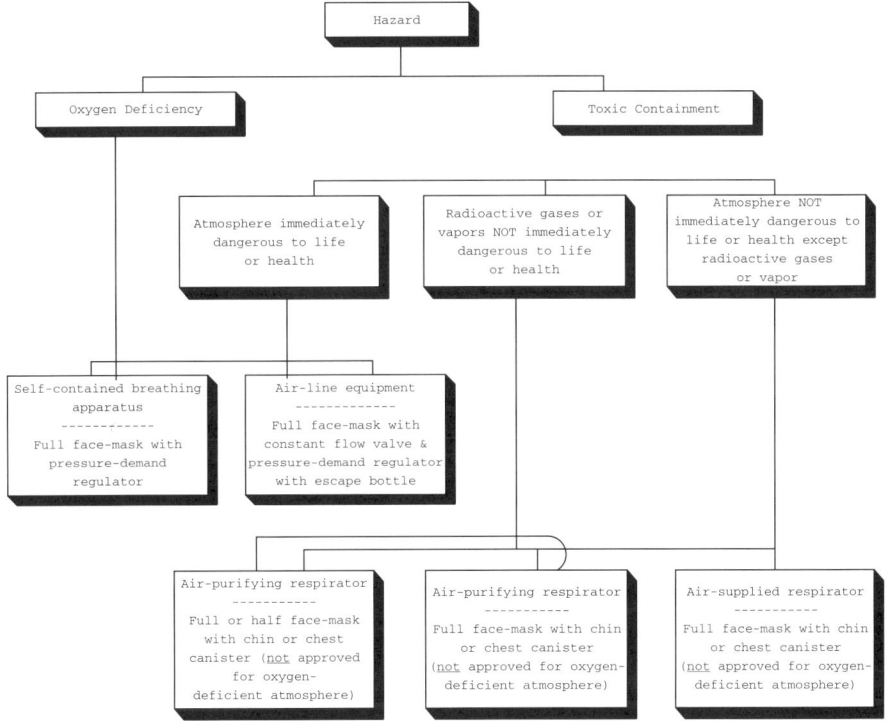

Figure 1.6   Example of respirator selection based on hazard identification.

THE EVALUATION PROCESS

After completing a preliminary survey, which is accomplished in a day or less, an environmental evaluation of worker exposure to airborne contaminants may be required. The evaluation can be made in two different ways:

1.   By sampling the worker's respiratory zone, usually accomplished by *personal exposure monitoring* (i.e., the worker is required to wear a personal sampling head and pump to determine the concentration of contaminants to which he is exposed) (see Figure 1.7).
2.   By sampling the workplace air, referred to as *area sampling*.

In recent years, personal exposure monitoring has become more popular because it gives the best estimate of worker exposure, taking into account parameters such as time and space fluctuations and different tasks performed during the shift. This method, in turn, gives the most reliable approach for assessing how much and what type of respiratory protection is required in a given circumstance.

The area sample, on the other hand, is collected to measure the amount of air contaminant to which the worker is exposed. The source of contamination may or

Figure 1.7   Personal air sampling (courtesy of 3M Canada).

may not be associated with what the people are doing in the area; however, if airborne concentration in an area is expected to remain fairly constant, then area samples may be used for estimating personal exposures. All jobs involving respirator wear should be regularly evaluated to ensure that the current respirator is still appropriate and adequate.

SAMPLING EQUIPMENT

Figure 1.8 shows the components of a personal exposure monitoring unit, which consist of the following:

- Sampling head as a collection device, such as a filter, tube, or impinger that includes a component that can trap or react to the contaminant being measured
- Flowmeter that indicates the rate of airflow through the sampler
- Pump or other suction device to draw environmental air

The following accessories are available with the aforementioned basic equipment:

1. *Pulsation damper*. Pulsation in the pump may interfere with proper functioning of the sampling intake, especially if the latter has, for instance, a

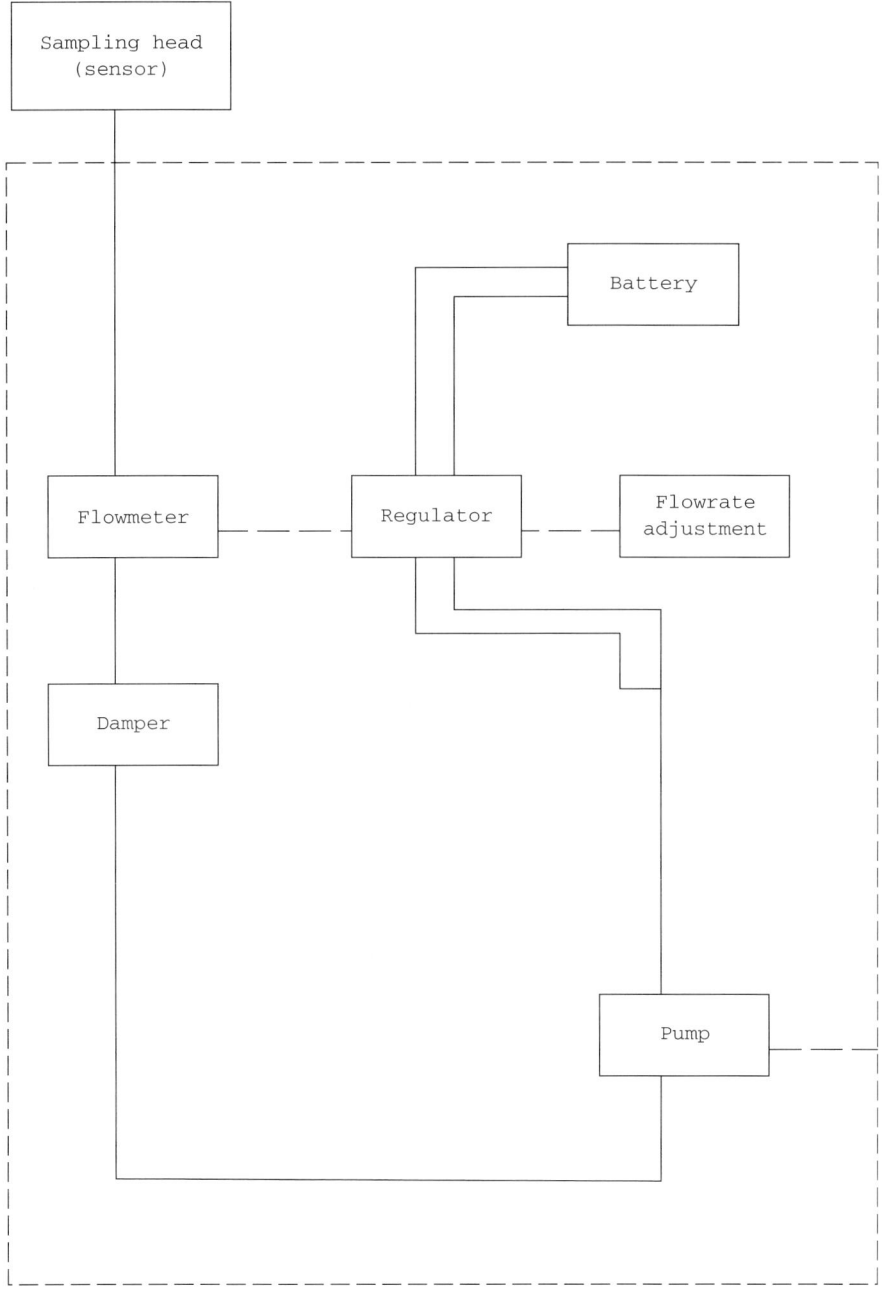

Figure 1.8   Components of a personal sampling unit.

particle-size selector cyclone. A more narrow chamber with flexible walls situated in series in the pump circuit acts as a capacity to damp pulsations in flowrate. This device has been incorporated into nearly all personal dust-sampling devices.

2. *Flowrate regulation devices.* The sampling of workplace air necessitates a relatively constant flowrate, especially for some types of particle-size selectors, the dimensions of which are calculated for a certain flowrate, and their efficiency is closely linked to a constant flowrate. Constant flowrate is also necessary to avoid gross errors in determining the volume sampled.

To overcome these problems, sampling instruments usually have a compensating device to limit the effects of variations in pressure upstream from the pump caused by progressive clogging of the filter.

Several types of direct-reading instruments are available to determine contaminant concentrations in air. Comprehensive information on these instruments is available in *Air Sampling Instruments for Evaluation of Atmospheric Contaminants*, published by the ACGIH.[3] These instruments read out the concentration of a collected sample directly without needing any elaborate laboratory analysis and feature detector tubes, direct-reading gas monitors, direct-reading dust monitors, and mercury vapor meters. The most popular among them are detector tubes.

A detector tube consists of a hand pump, a detector tube, and a calibrated scale or color standard. The detector tube is a sealed gas tube containing a solid support impregnated with a reactive chemical reagent. An air sample is drawn through the tube with the hand pump, and if a contaminant is present, it reacts with the reagent in the tube, producing a color. This color presents the level of concentration of the contaminant in the air.

In the range of 20% variation above or below the health standard, the tubes lack the accuracy required in making a decision. In this "gray area," samples should be taken using the sampling train described previously.

Another problem with detector tubes is the lack of specificity. Many tubes are far from specific, and hence, accurate knowledge of the possible interfering gases cannot be obtained. In order to quantify the effect of these interfering gases, and to avoid dangerously misleading results, the interpretation of detector tube measurements must be supervised by a qualified industrial hygienist.

Several battery-operated gas and vapor direct-reading instruments/monitors are available, including monitors for carbon monoxide, hydrogen sulfide, combustible gases, organic vapors, and oxygen. The theory, principles of operation, and application of these instruments are well documented in the literature.[3] The instruments are mainly grouped into the following classifications: electrochemical (operate on the principles of conductivity, potentiometry coulmetry, and ionization), spectrochemical (include infrared analyzers, ultraviolet and visible light photometers, chemiluminescent detectors, and photometric analysers), and thermochemical (conductivity or heat of combustion is measured).

These direct-reading instruments vary in performance depending on linear range, specificity, and limits of detection. Because they take "grab samples," they suffer from the same disadvantages as the detector tubes. It has been suggested that a series of grab samples may be used to determine the time-weighted averages over a long period; however, this involves careful planning regarding sampling strategy and the treatment of the sampling results obtained. Sampling results can, however, be helpful in determining the areas of high concentration and processes with the highest emissions.

Very few reliable direct-reading dust monitors are available, although considerable advances have been made in the last 10 years. They include light-scattering photometers, light-scattering single particle counters, light-attenuating photometers, and condensation nucleus counters. These instruments and their principles of operation are detailed in ACGIH's air sampling guide.[3] Different aerosol properties are measured by different direct-reading instruments, hence the properties may not be directly compared without some modifications of the data to account for the differences. Some of the properties determined include particle sizes, aerosol number concentrations, and aerosol mass concentrations. The accuracy of these instruments is generally limited because it depends on the relationship between the sensing region and aerosol property, and is based on an empirical formula using a "well-calibrated" aerosol system. The real aerosol measure may have a different, unknown relationship. A portable condensation particle counter is often used for respirator fit testing.

Another type of air-monitoring device is the diffusional monitor. It operates on the principle of diffusion, which is a process by which vapors and gases flow from high concentrations in the workplace to near-zero concentrations inside the monitor. The driving force for the sampling is the concentration gradient between the face of the monitor and the sorbent layer in the monitor. Air sampling with diffusional monitors provides a convenient and economical test of air quality in a worker's breathing zone. Diffusional monitors are easy to wear, simple to use, and highly accurate.[3]

SAMPLING PROCEDURE AND SAMPLING STRATEGY

Air sampling involves drawing air through a collection device at a known flowrate. The quantity of substance gathered on the device is then determined through laboratory analysis. From this information—and the known quantity of air that has been drawn through the sampler—it is possible to calculate the concentration of substances in the sampled air. In order to collect representative samples, it is essential to consider the following:

- Duration of sampling
- When to conduct the sampling

- Number of samples
- Frequency of sampling

The duration of sampling (i.e., the amount of time involved in drawing air through the sampling device) depends on the type of exposure concentration being measured. The recommended minimum sampling duration varies from contaminant to contaminant. When determining the time-weighted average exposure concentration, sampling is usually conducted for six to eight hours, or for a full work shift; however, a shorter sampling time may be used if it is planned to reflect representative exposure levels.

In most industrial operations, the concentration of contaminants in the air fluctuates. Before establishing a sampling strategy, it is important to have thoroughly evaluated the nature of the operation in order to determine when, and under what conditions, the substance may be emitted. Representative sampling times should then be chosen so that exposure over a full work week can be calculated. For example, if operations do not vary from day to day, it will probably be sufficient to base calculations of weekly exposure on full-shift sampling conducted for one day. If considerable daily variation exists, then sampling should be conducted over several days, representing the different exposure conditions. If exposure conditions vary from week to week, sampling should be performed during a week when maximum exposure is expected. The number of samples gathered should be planned to obtain a representative picture of exposure conditions. In order to compensate for errors inherent in sampling and analysis procedures, it is generally expected that a minimum of three samples will be gathered for each exposure situation. This may be done by taking separate samples, under similar exposure conditions, at the same time; it may also be done by sampling, in the same location, under the same exposure conditions, on different days.

The frequency with which air sampling is done is based on the exposure conditions in the individual workplace. If exposure levels are regularly near the allowable limits, air monitoring will be necessary more often. In such cases, monitoring may be appropriate on a monthly or quarterly basis. Where exposure levels are usually much lower than prescribed limits, semiannual monitoring should be sufficient; however, air sampling should always be performed when any changes in the process or conditions of exposure occur.

Sampling must be conducted by personnel (e.g., an industrial hygienist) who are trained to operate monitoring devices and accurately record essential information. If frequent monitoring is necessary, the employer may want to hire staff trained in sampling techniques or train employees who are already on staff. If this is not desirable, the employer may contract with private consultants to conduct air-monitoring tests or he may ask for assistance from a safety association. In most situations, employers will send the samples they have collected to a public or private chemical laboratory for analysis. In choosing a laboratory, the employer must ascertain that it will analyze the samples according to the standard methods and procedures.

STANDARDS AND GUIDELINES[4]

ACGIH

ACGIH exposure limits include three categories of threshold limit values (TLVs). They are as follows:

- *Time-weighted average (TWA)*. This exposure limit is based on acceptable contaminant concentrations for a normal, eight-hour workday and a 40-hour work week.
- *Short-term exposure limit (STEL)*. This is a 15-minute TWA exposure that should not be exceeded at any time during a workday. Exposures above the TWA up to the STEL should not be longer than 15 minutes and should not occur more than four times per day. Additionally, there should be at least 60 minutes between successive exposures in this range.
- *Ceiling (CEIL)*. This refers to concentrations that must not be exceeded during any part of the working exposure. As such, ceiling TLVs take precedent over all TWAs and STELs.

OSHA

OSHA mandates the following standards (www.OSHA.gov):

- *Permissible exposure limit (PEL)*. Based on an eight-hour TWA, PELs are exposure levels below which OSHA does not require respiratory protection. When exposures surpass this level, certain respiratory protection requirements must be met.
- *Short-term exposure limit (STEL)*. This is a 15-minute TWA exposure that should not be exceeded at any time during a workday.
- *Ceiling (CEIL)*. Same as ACGIH definition.

NIOSH

The National Institute for Occupational Safety and Health (NIOSH) recommends the following guidelines (www. cdc.gov/niosh):

- *Recommended exposure limit (REL)*. A TWA concentration for up to a 10-hour workday during a 40-hour work week.
- *Short-term exposure limit (STEL)*. Same as OSHA definition.
- *Ceiling (CEIL)*. Same as ACGIH definition.
- *Immediately dangerous to life or health (IDLH)*. As its name implies, the IDLH level refers to acute respiratory exposures that pose an immediate threat of loss of life. Exposures at this level may result in irreversible or severe

health effects, eye damage, irritation, or other conditions that could impair an employee's escape from the hazardous atmosphere.

AIHA

The American Industrial Hygiene Association (AIHA) exposure limits include the following workplace environmental exposure levels (WEELs):

- *Time-weighted average (TWA)*. Same as ACGIH definition.
- *Short-term exposure level (STEL)*. A 15-minute TWA.
- *Ceiling (CEIL)*. Same as ACGIH definition.

It is important to note that exposure limits and other exposure standards are constantly changing as more data is gathered about specific chemicals and substances.

Several countries have adopted the ACGIH values, giving them legal status in their health and safety regulations. In order to determine the time-weighted average exposure for comparison with the TLVs, samples that represent exposure conditions over the course of a day must be taken. The results of analysis give the airborne concentration of the substance during this time. From this information, the eight-hour time-weighted average exposure concentration can be determined by multiplying the concentrates C_1 by the time T_1 hours in which the worker is exposed to such concentration, and dividing the cumulative daily exposure by 8. This may be expressed as:

$$\frac{C_1 T_1 + C_2 T_2 + \ldots + C_n T_n}{8}$$

The basic difference between short-term exposures and ceiling values is that short-term levels are specified time-weighted average maximum or values for 15 minutes, whereas ceiling values are the maximum values at any time.

The air samples obtained for both short-term and ceiling values are usually taken for a much shorter period than those taken for calculating the time-weighted averages. Each measurement usually consists of a 15-minute sample (or series of sequential samples totaling 15 minutes) or an instantaneous air sample.

There may be occasions when the interday variations in exposure appear minimal because of the nature of the work process or operation. In such cases, additional measurements should be made to ensure that there will be one or more sampling periods during which the maximum concentrations can be detected.

Because no sample measurement gives an absolute, correct answer, and because every sample differs from the respective true average concentration, statistical methods are used to calculate interval limits for each side of the average-exposure concentration at a selected statistical confidence level (e.g., 95%). Two references have dealt with the details of statistical analysis.[5,6]

CONCLUSION

The respiratory tract is a sensitive system in the human body. It is assaulted daily by numerous contaminants that can cause harm to the respiratory tract itself or other vital organs of the body. It is essential that one be aware of the nature and levels of contaminants in the air. This chapter has summarized the different airborne hazardous materials and methods available to determine their concentration in the workplace.

REFERENCES

1. American Conference of Governmental Industrial Hygienists (ACGIH). *2001 TLVs and BEIs: Threshold Limit Values for Chemical Substances and Physical Agents and Biological Exposure Indices*. Cincinnati, OH: ACGIH, 2001.
2. Respiratory Protection Code of Federal Regulation, Title 29, Section 1910.134.1998.
3. American Conference of Governmental Industrial Hygienists (ACGIH). *Air Sampling Instruments for Evaluation of Atmospheric Contaminants*, 9th ed. Cincinnati, OH: ACGIH, 2001.
4. American Industrial Hygiene Association (AIHA). *The Occupational Environment: Its Evaluation and Control*, edited by S.R. DiNardi. Fairfax, VA: AIHA, 1997.
5. Rajhans, Gyan S., and D.S.L. Blackwell. *Practical Guide to Respirator Usage in Industry*. Boston: Butterworth–Heinemann, 1985.
6. Mulhausen, J.R., and J. Damiano. *A Strategy for Assessing and Managing Occupational Exposures*, 2nd ed. Fairfax, VA: AIHA, 1998.

2

Respirator Types, Uses, and Limitations

Workers can be exposed to many potential health hazards that may present an immediate or long-term threat to their health and well-being. Inhalation of toxic gases and airborne contaminants provides toxic materials the most direct access to the body and provides a means to distribute them to other areas of the body. As discussed in Chapter 1, there are two types of respiratory hazards—airborne contaminants and oxygen deficiency.

Airborne contaminants can take different forms, such as dust, mist, gas, vapor, and so on. Their effects on the human body can range from irritants to systemic poisons. These effects can be summarized into two categories:

1. *Acute effects*. These are immediate effects such as irritation of the respiratory tract, coughing, and poisoning. The person can notice such effects and take immediate precautionary measures.
2. *Chronic effects*. These effects occur after long-term exposure to low levels of airborne contaminants. The exposed person is unable to notice the presence of such effects immediately.

Under most occupational health and safety regulations, the health of workers must be protected by using engineering controls, and respirators should be used as a last resort. Typically, respirators are required under the following five conditions:

1. To reduce exposure during the time needed to install engineering controls.
2. To supplement engineering controls and work practices that fail to completely reduce the hazard to a permissible exposure level.
3. During activities such as maintenance and repairs, when engineering controls are not feasible.
4. During emergencies.
5. When measures and procedures necessary to control the exposure do not exist or are unavailable.

Engineering controls are by far the best method for effective and reliable protection from hazardous materials because they control the source of emission. Respirators, however, do not eliminate the hazard. Because of the five situations described previously, a larger worker population must rely on respirators for necessary protection.

In order to ensure proper protection for workers, the correct respirator must be selected, and the employer and worker must understand the limitations of each type of respirator. This chapter summarizes the different types of respirators available and the limitations of each type. Respirators protect the wearer either by removing the contaminant from the air before it is breathed in or by supplying a means of breathable air; therefore, respirators are divided into two broad categories:

1. *Air-purifying respirators*. These respirators remove contaminants from the air before it is breathed in (see Figures 2.1–2.6).
2. *Air-supplying respirators*. These respirators supply breathable air from an independent source.

Respirators are of two general "fit" types, tight-fitting and loose-fitting (see Figures 2.16–2.19).

Tight-fitting respirators are designed to form a seal with the face of the wearer (see Figure 2.1). They are available in three types:

1. *Quarter-mask*. Covers the nose and mouth, where the lower sealing surface rests between the chin and the mouth (see Figure 2.1).
2. *Half-mask*. Covers the nose and mouth and fits under the chin (see Figure 2.3).
3. *Full facepiece*. Covers the entire face from below the chin to the hairline (see Figure 2.2).

Loose-fitting respirators have a respiratory-inlet covering that is designed to form a partial seal with the face (see Figure 2.5).

These include loose-fitting facepieces, as well as hoods, helmets, blouses, or full suits, all of which cover the head completely. The best known loose-fitting respirator is the supplied air hood used by the abrasive blaster. The hood covers the head, neck, and upper torso, and usually includes a neck cuff. Air is delivered by a compressor through a hose leading into the hood. Because the hood is not tight-fitting, sufficient air must be provided to maintain a slight positive pressure inside the hood relative to the environment immediately outside the hood. In this way, an outward flow of air from the respirator prevents contaminants from entering the hood (see "Atmosphere-Supplying Respirators" section).

AIR-PURIFYING RESPIRATORS

Air-purifying respirators (Figures 2.1 through 2.5) are designed to remove contaminants in the form of particulates, gases, and vapors, or combinations of all of

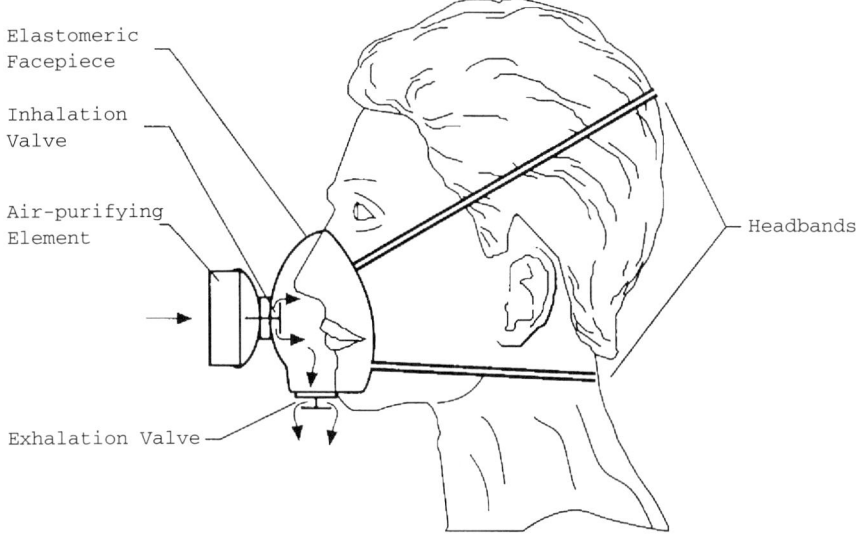

Elastomeric
Facepiece

Inhalation
Valve

Air-purifying
Element

Headbands

Exhalation Valve

Figure 2.1 Reusable air-purifying respirator, nonpowered, equipped with quarter-mask type elastomeric face piece. (Reprinted with permission from *Respirator Protection Handbook*, Lewis Publishers, CRC Press, Boca Raton, Florida[1].)

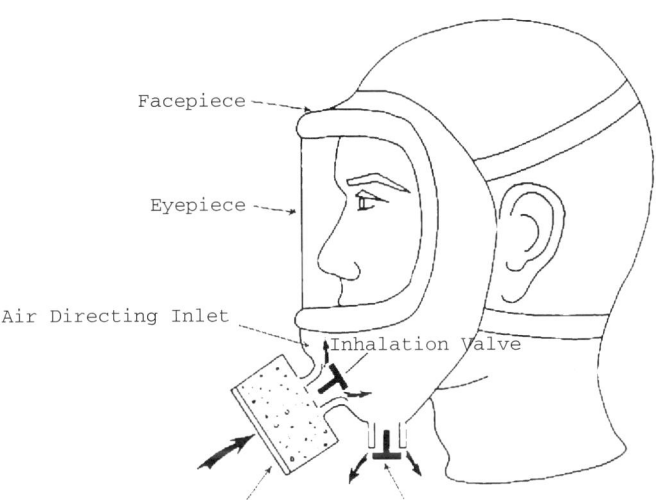

Facepiece

Eyepiece

Air Directing Inlet

Inhalation Valve

Figure 2.2 Nonpowered air-purifying respirator equipped with full facepiece respiratory-inlet covering the canister-type air-purifying element. (Reprinted with permission from *Respirator Protection Handbook*, Lewis Publishers, CRC Press, Boca Raton, Florida.)

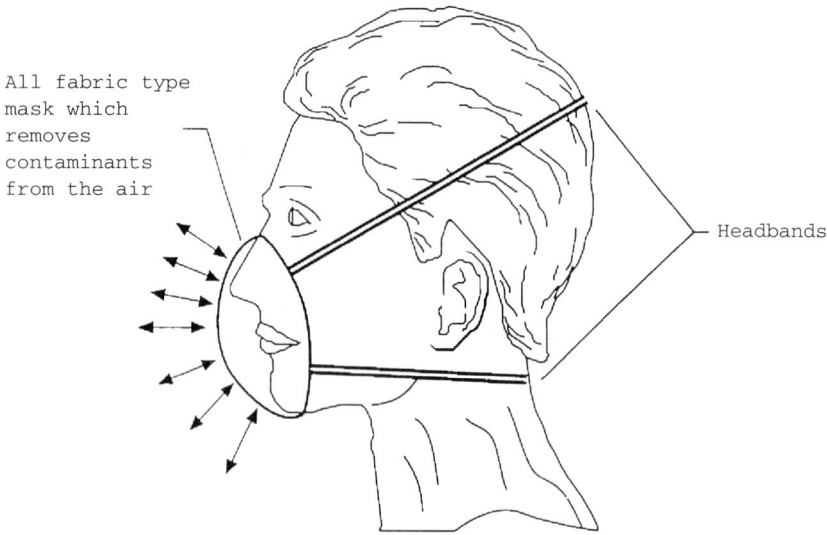

All fabric type
mask which
removes
contaminants
from the air

Headbands

Figure 2.3   Disposable nonpowered air-purifying respirator with all-fabric mask (valveless) half-mask type. (Reprinted with permission from *Respirator Protection Handbook*, Lewis Publishers, CRC Press, Boca Raton, Florida.)

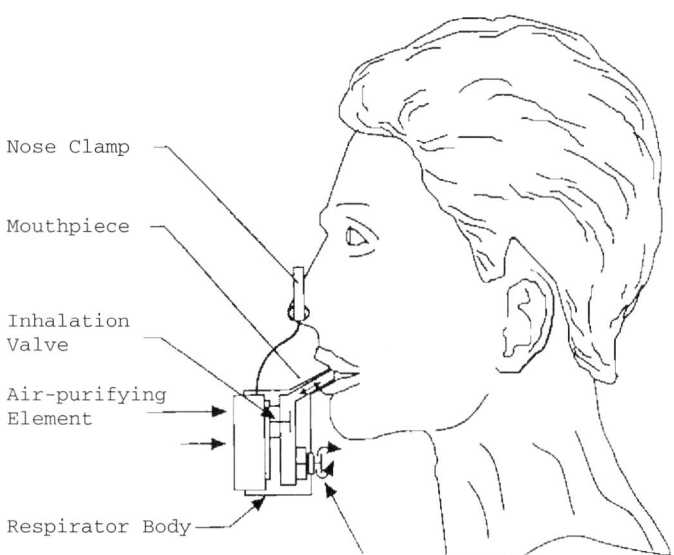

Nose Clamp

Mouthpiece

Inhalation
Valve

Air-purifying
Element

Respirator Body

Figure 2.4   Nonpowered air-purifying respirator equipped with a mouthpiece and nose-clamp type respiratory-inlet covering. (Reprinted with permission from *Respirator Protection Handbook*, Lewis Publishers, CRC Press, Boca Raton, Florida.)

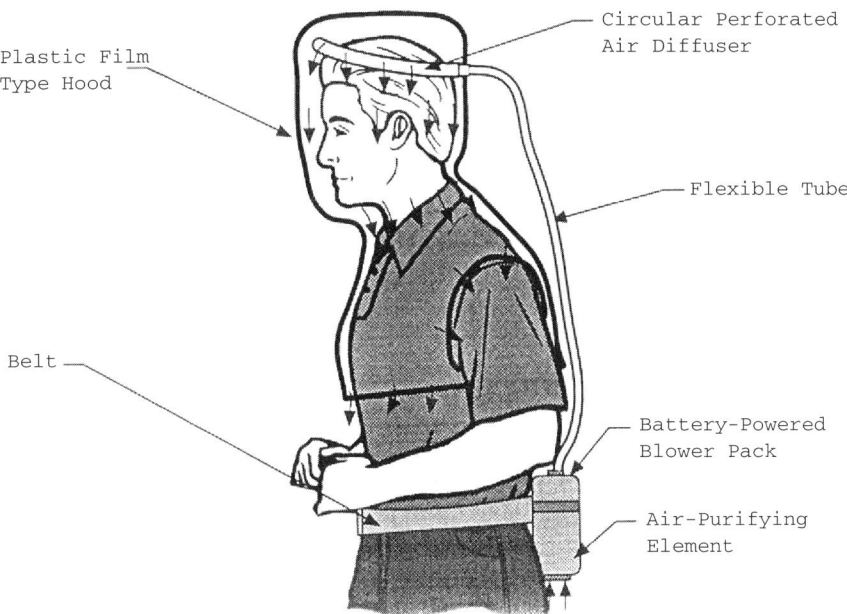

Figure 2.5 Powered air-purifying respirator equipped with hood-type loose-fitting respiratory-inlet covering. (Reprinted with permission from *Respirator Protection Handbook*, Lewis Publishers, CRC Press, Boca Raton, Florida.)

these. The composition of the filtering medium depends on the contaminant to be filtered: various chemicals are used to remove specific gases and vapors, whereas mechanical filters remove particulate matter. Within this classification, there are four basic types of air-purifying devices:

1. Mechanical-filter (particulate) respirator
2. Chemical-cartridge respirator
3. Gas mask
4. Powered air-purifying respirator

Mechanical-Filter Respirators

The mechanical-filter respirator provides respiratory protection against particulate matter such as dust, mist, and fumes. Usually, a fibrous material is used to trap the particulates. The efficiency of a filter medium depends on the size of the particle relative to the filter size, particle velocity, and, to some extent, the composition and shape of both the particle and fiber. Figure 2.6 shows different types of mechanical-filter respirators.

Figure 2.6   Different types of mechanical-filter respirators (courtesy of 3M Canada).

The mechanism of filtration can be either simple straining or depth filtration. Straining depends on the solid particles being collected because they are larger than the pores in the medium. Depth filtration depends on the particles adhering to the exposed surface of the filter medium within its depth. This type of filtration operates on a combination of mechanical and electrical forces: interception, impaction, diffusion, and electrostatic. The specific type of force depends mainly on the particle size.

As air flows through a filter, the presence of fibers results in a curvature of the streamlines in the vicinity of the fiber. Very small particles—less then 0.01 microns in size—will not be carried by the air current. Brownian motion, or diffusion, controls their movement, and eventually this random movement results in a collision with a solid surface. With decreasing particle size, diffusion becomes more intense and, as a result, the intensity of diffusional deposition increases. This mechanism is most significant in removing fine dusts. Larger particles may be trapped either through impaction or interception. Impaction occurs when a particle, because of its momentum, crosses the streamlines and impinges on the fiber. The frequency of impaction increases with increasing particle size, and the velocity of flow, and it may also be referred to as *inertial interception*. Direct interception occurs when a particle is intercepted as it approaches the collecting surface at a distance equal to its radius. Figure 2.7 illustrates each of these filtering forces.

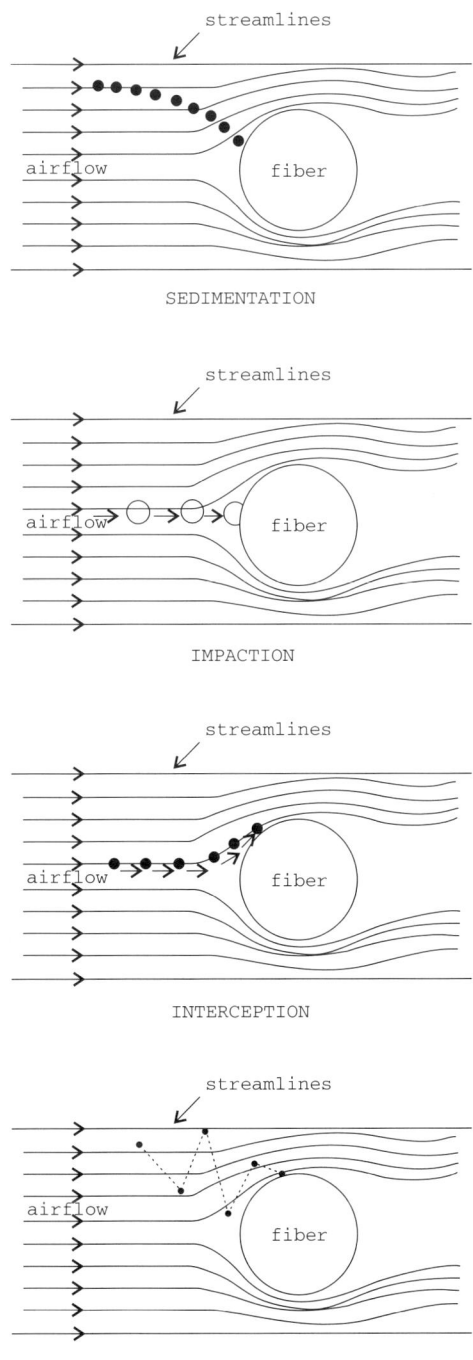

Figure 2.7 Filtration mechanisms.

Electrostatic action is extremely important for removing small particles. Fibers and particles have electrical charges. Generally, increasing the charges on either of the two increases the opportunity for collecting the particle. The electrical charge generates a force that helps the particle move from the air streamline to a nearby fiber. A particle can move more than one radius across the air stream to a charged fiber (Figure 2.8). The fibers of a filter can be charged to influence particle collection. The wool resin filter, in which an electrostatic charge is developed when the wool aid resin are carded together on a clothing machine, has been available since World War II. The main disadvantage of this filter medium is that the charge on the fibers is not always stable and decreases with time.

Another development is the electret-fibrous filter. An electret is plastic with electrostatic charges physically embedded into the surface, which is then formed into a fibrous blanket. They can be seen as the electrical counterparts of magnets (i.e., they consist of dielectrics carrying a strong positive and negative electric charge).

The advantage of electrical forces over mechanical ones in trapping particulates is that electrical forces are active at some distance from the fibers. As a result, filters made from electrets have a rather open structure with large spaces, even if minute, submicron particles are captured efficiently. The open structure allows a stream of dust-laden air to pass through easily so that the air resistance of electret filters is low, even at fairly high flowrates.

The mechanical-filter or particulate-removing respirator group is generally called dust, mist, or fume respirators, or combinations thereof. All dust, mist, and fume respirators protect in exactly the same way, by removing and retaining the airborne particulate before it can be inhaled. The mechanical-filter itself is generally composed of wool or a synthetic fiber. It may also consist of a filter composed of a resin-impregnated blend to which an electrostatic charge is imparted in order to increase its efficiency by electrostatically attracting the particles to the fibers.

No filter is 100% efficient in removing particles. The probability that a single particle will be trapped depends on many factors, including the composition and

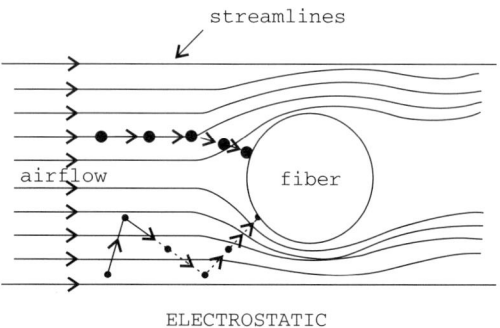

Figure 2.8   Electrostatic filtration mechanism.

shape of both the particle and the fiber. Hence, mechanical filters are also classified as designed for protection against dust, mists, fumes, and any combination thereof. For example, "toxic dust" filters are designed for protection against toxic dusts whose threshold limit value (TLV) is $0.05 \, \text{mg/m}^3$ or more. Some of these filters are also approved for mists whose TLV is $0.05 \, \text{mg/m}^3$ or more. "Fume" filters are approved for contaminants, such as metal fumes, whose TLV is $0.05 \, \text{mg/m}^3$ or more, but have a higher efficiency toward particles less than 1 μm in diameter. Fume filters are generally accepted as being 99.5% efficient against 0.6 μm particles as opposed to 99.0% for dust filters. High-efficiency filters are designed to protect against particulate contaminants with a TLV less than $0.05 \, \text{mg/m}^3$ and are at least 99.97% efficient against 0.3 μm particles. High-efficiency filters, however, have poorer particle-loading characteristics, which tend to increase breathing resistance.

There are two basic designs of mechanical-filter respirators: cartridge and disposable (Figure 2.9). The cartridge respirator consists of a molded rubber or plastic facepiece, which, when placed on the face, makes a seal between the skin and the molded facepiece. The cartridges, which act as the filtering medium, are screwed onto the facepiece. Cartridges are available for protection against particulates, gases/vapors, or combinations of the two. The cartridges are replaced when they are used up. In contrast to the cartridge-type respirator, disposable respirators have nonreplaceable parts. The entire respirator is disposed of when it is no longer usable. This option eliminates the need for a maintenance program.

Depending on approval from the National Institute of Occupational Safety and Health (NIOSH) [DHHS (NIOSH) Publication No. 87-116],[2] disposable respirators may be considered as *single-use* or *disposable*. A single-use respirator refers to a specific certification granted by NIOSH. A disposable respirator can have NIOSH certification for single-use, dust, mist, fume, high-efficiency, or chemical-cartridge. Single-use and disposable are not equivalent descriptive terms for respiratory equipment. Single-use refers to respirators approved by NIOSH for use against lung-damaging dusts only, and must meet the requirements as specified in Federal Register 42CFR84.[3] Replaceable refers to respirators that can be reused (i.e., they can be removed and put back on and still will provide the necessary protection). These replaceable respirators are tested and must pass the requirements outlined in 42CFR84 (also referred to as Part 84). Both the cartridge-type and the disposable-type provide the same protection because both are considered replaceable and pass the same NIOSH test criteria (summarized in Table 2.1).

Federal Register 42CFR84 replaces the old certification standard 30CFR11.[4] Some of the particulate/respirator certification tests in the old standard were based on Bureau Mines procedures used in 1930s and were never significantly updated. NIOSH recognized that new research, testing, and manufacturing technology had rendered the particulate filter/respirator testing procedures in the old standard obsolete.

NIOSH established the new test criteria to simulate worst-case respirator use and severe test conditions. These filters can be used without particle-size analysis or filter penetration testing in the workplace. R- or P-series filters should be

(a)

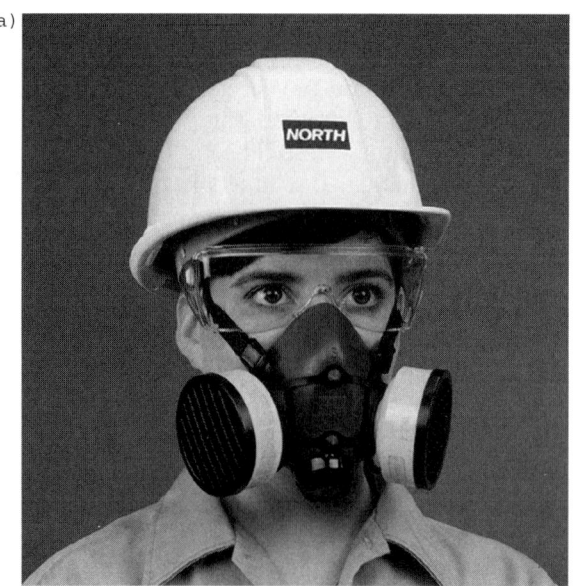

(b)

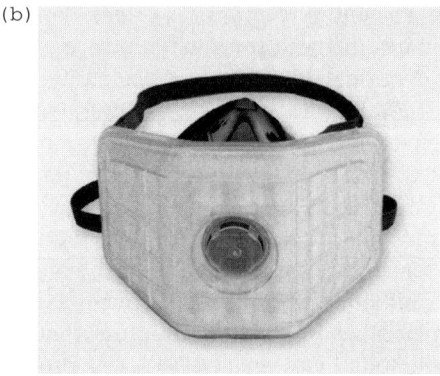

Figure 2.9   (a), Cartridge-mechanical-filter respirator (courtesy of North Safety Products); (b), Disposable mechanical-filter respirator (courtesy of North Safety Products).

selected if there are oil (e.g., lubricants, cutting fluids, glycerine) or nonoil aerosols in the workplace. N-series filters should be used only for nonoil (i.e., solid and water-based) aerosols.

Note: To help you remember the filter series, use the following guide:

N for **N**ot resistant to oil
R for **R**esistant to oil
P for oil **P**roof

Table 2.1   42CFR84 filter classes[3]

CLASS OF FILTER	EFFICIENCY (%)	TEST AGENT	TEST MAXIMUM LOADING (MG)	TYPE OF CONTAMINANT	SERVICE TIME[*]
N-series	—	NaCl[†]	200	Solid and water-based particulates (i.e., nonoil aerosols)	Nonspecific[†, §]
N 100	99.7				
N 99	99				
N 95	95				
R-series	—	DOP oil[**]	200	Any	One work shift[†, ††]
R 100	99.7				
R 99	99				
R 95	95				
P-series	—	DOP oil		Stabilized any efficiency	Nonspecific[‡]
P 100[‡‡]	99.7				
P 99	99				
P 95	95				

[*]NIOSH will be conducting and encouraging other researchers to conduct studies to assure that these service time recommendations are adequate. If research indicates the need, NIOSH may recommend additional service time limitations for specific workplace conditions.
[†]NaCl = sodium chloride.
[‡]Limited by considerations of hygiene, damage, and breathing resistance.
[§]High (200 mg) filter loading in the certification test is intended to address the potential for filter efficiency degradation by solid or water-based (i.e., nonoil) aerosols in the workplace. Accordingly, there is no recommended service time limit in most workplace settings; however, in dirty workplaces (high aerosol concentrations), service time should be extended beyond 8 hours of use (continuous or intermittent) only by performing an evaluation in specific workplace settings that: (1) demonstrates that extended use will not degrade the filter efficiency below the certified efficiency level, or (2) demonstrates that the total mass loading of the filter is less than 200 mg (100 mg per filter for dual-filter respirators).
[**]DOP oil = dioctyl phthalate.
[††]No specific service time limit is needed when oil aerosols are not present. In the presence of oil aerosols, service time may be extended beyond 8 hours of use (continuous or intermittent) by (1) demonstrating that extended use will not degrade the filter efficiency below the certified efficiency level, or (2) demonstrating that the total mass loading of the filter is less than 200 mg (100 mg per filter for dual-filter respirators).
[‡‡]The P100 filter must be color-coded magenta. The Part 84 Subpart KK HEPA filter on a PAPR will also be magenta, but the label will be different from the P100 filter, and the two filters cannot be interchanged.

The filter certification test is called worst-case (i.e., it produces maximum filter penetration) because the test conditions are the most severe that are likely to be encountered in a work environment. These conditions are:

- Air flow that simulates a high work rate (85 + 4 liters per minute for single filters; 42.5 + 2 liters per minute through each filter for paired filters)
- The most penetrating aerosol size (approximately 0.3 micrometer)
- Charge-neutralized particles
- The most filter-degrading test aerosol for R- and P-series filters
- Measurement of instantaneous (not average) penetration
- High total filter loading (up to 200 mg for N- and R-series filters, and continued loading until there is no further decrease in efficiency for P filters)

The degradation categories (N-, R-, and P-series) are determined by using either sodium chloride (NaCl) or dioctyl phthalate (DOP) as the test aerosol. NaCl is only slightly degrading to filter efficiency, whereas DOP is very degrading. Respirators tested with NaCl (i.e., N-series filters) are not resistant to efficiency degradation by oils and should be used only in workplaces free of oil aerosols. Filters passing DOP oil tests (i.e., R- and P-series filters) are resistant to efficiency degradation and can be used to protect against any aerosols (including oil-based particulates) in the workplace.

USE LIMITATIONS

The service-life of all three categories of filter efficiency degradation (i.e., N-, R-, and P-series) is limited by considerations of hygiene, damage, and breathing resistance. All filters should be replaced whenever they are damaged, soiled, or causing noticeably increased breathing resistance (e.g., causing discomfort to the wearer).

R- or P-series filters can be used to protect against oil or nonoil aerosols. N-series filters should be used only for nonoil aerosols. Use and reuse of the P-series filters would be subject only to considerations of hygiene, damage, and increased breathing resistance. Generally, the use and reuse of N-series filters would also be subject only to considerations of hygiene, damage, and increased breathing resistance; however, for dirty workplaces that could result in high filter loading (i.e., 200 mg), service time for N-series filters should be extended beyond 8 hours of use (continuous or intermittent) by performing an evaluation in specific workplace settings that: (1) demonstrates that extended use will not degrade the filter efficiency below the efficiency level specified in Part 84, or (2) demonstrates that the total mass loading of the filter(s) is less than 200 mg. The R-series filters should be used only for a single shift (or for 8 hours of continuous or intermittent use) when oil is present. Service time for the R-series filters can be extended using the same two methods described for N-series filters. These determinations would need to be repeated whenever conditions change or modifications are made to processes that could change the type of particulate generated in the user's facility.

The new NIOSH approval number TC-84A-XXX is not required to be shown on the filter/respirator. This is in contrast to the old approval, which required the NIOSH approval number TC-21C-XXX to be shown on the filter/respirator. The only information required to be placed and usually shown on the filter/respirator is the NIOSH classification (e.g., N 95, R 99, P 100).

There are two criteria used for replacing a dust/mist/fume respirator, or combinations thereof: (1) the respirator or filter medium is damaged, or (2) the filter or respirator is difficult to breathe through. Damage to the respirator or filter is usually evident, and thus the wearer should be trained to easily determine that the respirator or filter must be replaced.

The second criterion is much more difficult to determine. As the inhaled air is filtered, the particles build up in the filter medium and clog the pores. This results in increasing breathing resistance. Eventually, resistance reaches a point where the wearer finds breathing uncomfortable and the filter must be discarded. The concentration of dust being filtered, the breathing rate of the wearer, the tolerance of the wearer, the loading characteristic of the filter, and the breathing resistance of the filter before loading influence the length of time required to reach an uncomfortable breathing resistance.

There is no accurate method for predetermining the service-life of a particulate respirator. The concentration of airborne dust is the major factor influencing service-life. The more dust in the air, the more dust the filter must be able to retain, and the faster it will clog. The breathing rate of the wearer will also influence the service-life. If a worker is doing light work, his breathing rate is approximately 20 liters per minute. If the workload is heavy, the breathing rate may be 60 liters per minute or higher. This results in approximately three times the volume of air going through the filter and thus three times as much dust to be filtered and retained in the filter.

Mechanical-filter respirators are usually available in three different categories: quarter-face, half-face, and full-face (Figure 2.10). The quarter-face covers the mouth and nose; the half-face covers the mouth, nose, and chin; and the full-face respirator covers the face from the chin to the hairline and from ear to ear.

Proper selection of the correct mechanical-filter respirator affords the worker with protection from nuisance dust, toxic dusts, mists, fumes, or any combinations, but there are important limitations:

● They do not provide oxygen, so they must not be used in oxygen-deficient atmospheres.
● They provide no protection against gases or vapors.
● There is a pressure drop through the filter medium; therefore, there is some breathing resistance. As the medium loads up, this resistance increases; thus, people with respiratory problems should ensure that they are capable of wearing this type of respirator.

Chemical-Cartridge Respirators

Chemical-cartridge respirators are vapor and gas removing, using a cartridge attached to the facepiece containing chemicals to trap or react with specific vapors or gases and remove them from the air breathed (Figure 2.11). The filtering medium is commonly activated carbon, which primarily removes organic vapors. Activated carbon can also be impregnated with other substances to make it more selective against specific gases and vapors. For example, iodine is added to absorb mercury vapors, metallic oxides for acid gases, and salts of metals are added to remove ammonia.

Similar to mechanical-filter respirators, several factors influence the service-life of a chemical-cartridge respirator: breathing rate, humidity, contaminant concentration, temperature, and so on. These factors make it difficult to give a general statement about how long a specific gas/vapor respirator will last.

(a)

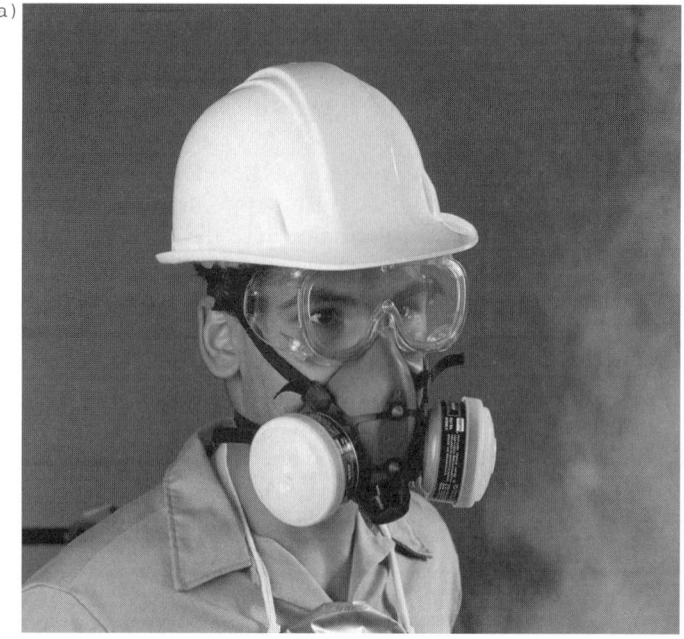

(b)

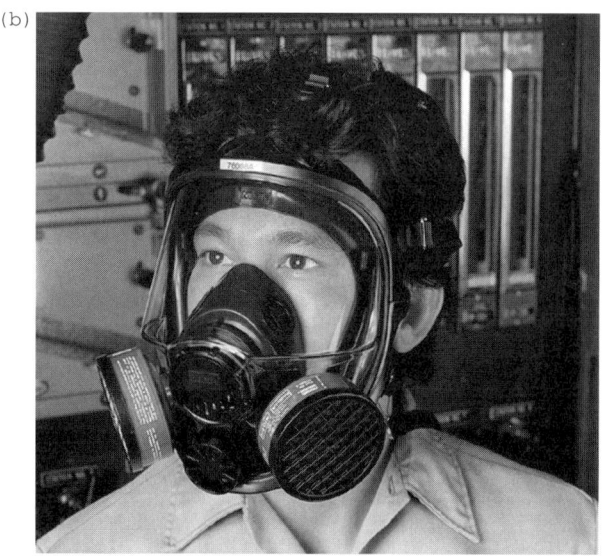

Figure 2.10 Half-face (a) and full-face (b) mechanical filter respirators (courtesy of North Safety Products).

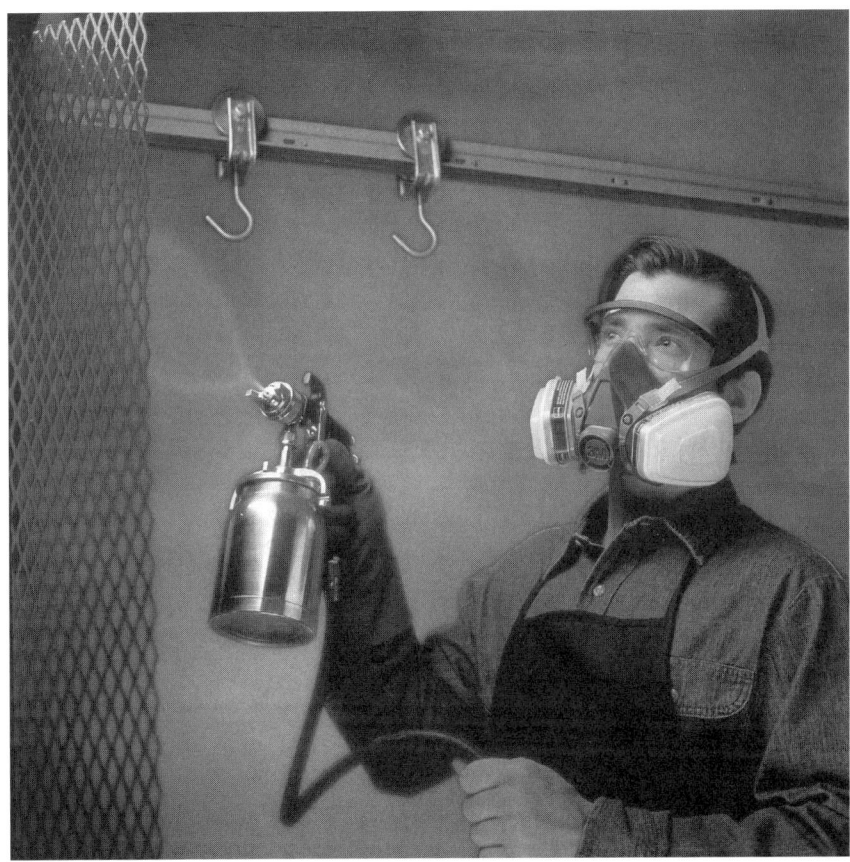

Figure 2.11 Chemical-cartridge respirator (courtesy of 3M Canada).

In order to be NIOSH-approved, a gas/vapor respirator must pass several service-life tests, which are outlined in 42CFR84. Table 2.2 summarizes these test requirements. One of the test requirements is a minimum service-life, which is determined when the permitted penetration value is detected.

Humidity, organic vapor volatility, concentration, carbon weights, contaminant molecular weight, and breathing rate all play a role in influencing cartridge service-life. Solvent type, flowrate, solvent concentrations, and relative humidity are the four major influencing factors in determining cartridge service-life. In addition, it should be noted that:

1. Each organic vapor has different adsorption characteristics on carbon. In general, it is believed that the higher the volatility of the vapor, the less efficiently it adsorbs into the carbon.
2. The more carbon available, the longer the service-life is. Doubling the amount of carbon will generally double the service-life.

Table 2.2   Summary of NIOSH test requirements for chemical-cartridge respirators

		TEST ATMOSPHERE				
CARTRIDGE	TEST CONDITION	GAS OR VAPOR	CONCEN-TRATION (PPM)	FLOWRATE (LITERS PER MINUTE)	PENE TRATION (PPM)	MIN. LIFE (MIN)
Ammonia	As received	NH_3	1000	64	50	50
	Equilibrated	NH_3	1000	32	50	50
Chlorine	As received	Cl_2	500	64	5	35
	Equilibrated	Cl_2	500	32	5	35
Hydrogen Chloride	As received	HCl	500	64	5	50
	Equilibrated	HCl	500	32	5	50
Methyl amine	As received	CH_3NH_3	1000	64	10	25
	Equilibrated	CH_3NH_3	1000	32	10	25
Organic vapors	As received	CCl_4	1000	64	5	50
	Equilibrated	CCl_4	1000	32	5	50
Sulfur dioxide	As received	SO_3	500	64	5	30
	Equilibrated	SO_3	500	32	5	30

Maximum Resistance
(mm water-column height)

	INHALATION		EXHALATION
TYPE OF CHEMICAL-CARTRIDGE RESPIRATOR	INITIAL	FINAL	
For gases, vapors, or gases and vapors	40	45	20
For gases, vapors, or gases and vapors, and dusts, fumes, and mists	50	70	20

3. Breakthrough time is inversely proportional to the flowrate (i.e., halving the flowrate will double the service-life).
4. The service-life is inversely proportional to the log of the concentration.
5. Service-life is shorter if the relative humidity is greater than 65%. If the humidity is less than 50%, the service-life is not noticeably affected.

It is not an easy matter to determine the actual service-life of a gas/vapor respirator because of all the variables that have influence. The best policy is to replace the respirator or cartridge when (1) an odor or taste is detected (see Table 2.3); (2) it becomes difficult to breath through; or (3) the cartridge or respirator is damaged.[3]

A safety factor should be built in. It is recommended that a 50% safety factor be applied when predicting the service-life of a chemical-cartridge respirator. The final judgment of the service-life should be with the wearer.

The effectiveness of gas and vapor respirators depends on their ability to remove chemical molecules from the air stream. Molecules of gas and vapor are

Table 2.3　Odor threshold for selected contaminants

COMPOUND	ODOR THRESHOLD (PPM/V)	COMPOUND	ODOR THRESHOLD (PPM/V)
Acetaldehyde	$0.031 - 0.21$	Ethyl metacrylate	0.0067
Acetic acid	$1.0 - 24.0$	Ethyl pelargonate	0.0014
Acetone	$40.0 - 100.0$	Ethyl selenide	0.0012
Acetophenone	$0.17 - 17.0$	Ethyl selenomer-captan	$3 \times 10^{-4} - 12 \times 10^{-3}$
Acrolein	$1.1 \times 10^{-5} - 1.5$	Ethyl sulphide	$6 \times 10^{-4} - 0.002$
Acrylonitrile	$19.0 - 21.4$	Ethyl i-valerate	0.12
Allyl alcohol	$0.017 - 1.4$	Ethyl n-valerate	0.060
Allyl amine	6.3	Ethylene	$4 \times 10^{-3} - 260$
Allyl chloride	$1.5 \times 10^{-4} - 0.47$	Ethylene bromide	25.0
Allyl isocyanide	0.018	Ethylene glycol	0.080
Allyl isothiocyanide	$0.0081.1 \times 10^{-5} - 1.50.15$	Ethylene oxide	$260.0 - 700.0$
Allyl mercaptan	$5 \times 10^{-5} - 0.004$	Ethylene undecanoate	5.4×10^{-4}
Allyl sulphide	1.4×10^{-4}	Eugenol	0.0046
Ammonia	$34.0 - 55.0$	Formaldehyde	$0.98 - 49.9$
n-Amyl acetate	$0.005 - 0.08$	Formic acid	$0.025 - 21.0$
Amyl alcohol	$0.12 - 35.0$	n-Heptane	22.0
Amylene	$0.0022 - 2.1$	Hexanol	$0.0050 - 5.2$
i-Amyl mercaptan	0.0043	Hydrazine	3.0
Amyl sulphide	0.003	Hydrocyanic acid	$2.7 \times 10^{-4} - 0.001$
Anethole	0.003	Hydrogen chloride	10.0
Aniline	$1.0 - 70.0$	Hydrogen sulphide	$0.0081 - 0.12$
Apiol	0.0063	Iodoform	5×10^{-4}
Benzene	$4.68 - 31.0$	Light gasoline	800.0
Benzyl chloride	0.04	Menthol	1.5
Benzyl mercaptan	0.0026	2-Mercato-ethanol	0.64
Benzyl sulfide	$0.002 - 0.006$	Mesitylene	0.027
Bromine	$0.047 - 1.0$	Methanol	$4.3 - 5900.0$
Bromoacetone	0.090	Methoxynaphthalene	1.2×10^{-4}

Table 2.3   *(Continued)*

COMPOUND	ODOR THRESHOLD (PPM/V)	COMPOUND	ODOR THRESHOLD (PPM/V)
Butadine	0.16 – 0.45	Methyl acetate	200.0
i-Butane	1.2	Methyl amine	0.02 – 3.3
n-Butanol	5.0 – 11.0	2-Methyl – 2-Butynate	2.3
s-Butanol	43.0	Methyl n-Butynate	0.0026
t-Butanol	73.0	Methyl chloride	10.0
2-Butanone	2.0 – 80.0	Methyl dichloro arsine	0.11
Butyl acetate	20.0	Methyl ethyl pyridine	0.05
Butyl formate	17.0	Methyl formate	2000.0
Butyl mercaptan	0.006	Methyl glycol	60.0
Butyl sulphide	0.015	Methyl isobutyl ketone	0.1 – 3.0
Butyric acid	5.6×10^{-4} – 0.4	Methyl mercaptan	0.041
Camphor	1.2 – 1.6	Methyl metacrylate	0.051 – 0.21
Carbon disulphide	0.081 – 0.02	2-Methyl propane	0.57
Carbon monoxide	odorless	Methyl salicylate	0.006
Carbon tetrachloride	21.0 – 200.0	Methyl sulphide	0.00015 – 0.012
Chloral	0.047	Methyl thiocyanate	0.25
Chlorine	0.01 – 0.3	Methyl vinyl pyridine	0.04
Chlorine dioxide	0.1	Methylene chloride	150.0 – 210.0
Chlorobenzene	0.21 – 94.0	Musk (synthetic)	4×10^{-7} – 0.004
Chloroform	200.0	Napthalene	0.027 – 6.8
m-Cresol	0.01 – 0.68	2-Napthol	1.3
o-Cresol	0.09 – 0.68	Nitric oxide	1.0
Crotonaldehyde	0.035 – 1.6	Nitorbenzene	0.0047 – 1.9
Cyclohexane	0.41	n-Octane	150.0
o-Dichlorobenzene	50.0	Octanoic acid	0.0014 – 35.0
1,1-Dichloroethane	120.0	1-Octanol	0.0021 – 0.1
1,2-Dichloropropane	50.0	2-Octanol	0.0026

COMPOUND	ODOR THRESHOLD (PPM/V)	COMPOUND	ODOR THRESHOLD (PPM/V)
Diethylamine	0.006 – 30.0	Oenathic acid	0.015
Diethyl selenide	1.4×10^{-4}	Ozone	0.02
Diethyl sulphide	0.004	Pelargonic acid	$8.6 \times 10^{-4} - 5.0$
Dimethyl acetamide	46.8	n-Pentane	2.2
Dimethyl formamide	100.0	Pentanol	0.0065 – 1.0
1,1-Dimethyl hydrazine	6.0	Pentanone	8.0 – 7.0
Dimethyl sulphide	0.02	i-Pentyl acetate	0.0028
Dioxane	0.8 – 170.0	n-Pentyl acetate	9.0×10^{-4}
Diphenyl chlorasine	0.03	Propyl mercaptan	0.0016
Diphenyl cyanoarsine	0.3	Propyl sulphide	0.011
Diphenylether	0.1	Pyridine	0.03 – 0.82
Diphenyl sulphide	0.0021 – 0.0047	Styrene	0.047 – 37.0
Diphosgene	1.2	Sulphur dioxide	0.47 – 3.0
Dithioethylene glycol	0.0031	Tetrachlorethylene	4.6 – 50.0
Dodecanol	0.0064	Tetrahydrofuran	30.0
Ethane	89.9 – 150.0	Toluene	1.7 – 40.0
1,2-Ethanedithiol	0.0031 – 0.0042	Toluene-2,4 disolcyanite	2.1
Ethanol	0.1 – 5100.0	Trichloroethylene	21.0 – 400.0
Ethyl acetate	0.056	Trimethylamine	$2.1 \times 10^{-4} - 1.7$
Ethyl acrylate	0.00047	Valeric acid	6×10^{-4}
Ethyl butyrate	0.0082 – 0.015	Vanillin	$3.2 \times 10^{-8} - 0.68$
Ethyl decanoate	0.00017	o-Xylene	20
Ethyl dichloroarsine	104	m-Xylene	3.7
Ethyl ether	0.33	p-Xylene	0.49
Ethyl glycol	25	Xylidene	0.0048
Ethyl mercaptan	$1.9 \times 10^{-4} - 5.1 \times 10^{-4}$		

much smaller than aerosol particles and are not removed by fiber filters. It is necessary to bring the molecules in contact with the sorbent material, which reacts with the contaminant. The reaction may be one of chemical combining (KOH + HCl) or of physical chemistry force (organic vapor and charcoal). Absorption and adsorption both may occur. Absorption is penetration; adsorption is adherence. The chemical reaction, absorption or adsorption process, is 100% efficient until the sorbent's capacity to filter out the contaminant is exhausted. At that time, the contaminant will pass completely through the sorbent material and into the facepiece. The *breakthrough* of the gas or vapor is generally the only way the wearer knows that the respirator is used up. Therefore, chemical-cartridge respirators can only be used against contaminants when good warning properties are employed (i.e., those that can be tasted or smelled at concentrations that are not toxic).

When particulates, gases, and vapors are present in the same environment, particulate prefitters can be placed in front of the chemical cartridge, thus providing a combination particulate/gas and vapor respirator (Figure 2.12).

Similar to the mechanical-filter respirator, the chemical-cartridge respirator is available as cartridge-type or disposable. It is also available in quarter, half, and full-face designs. Users must be aware of the following limitations to

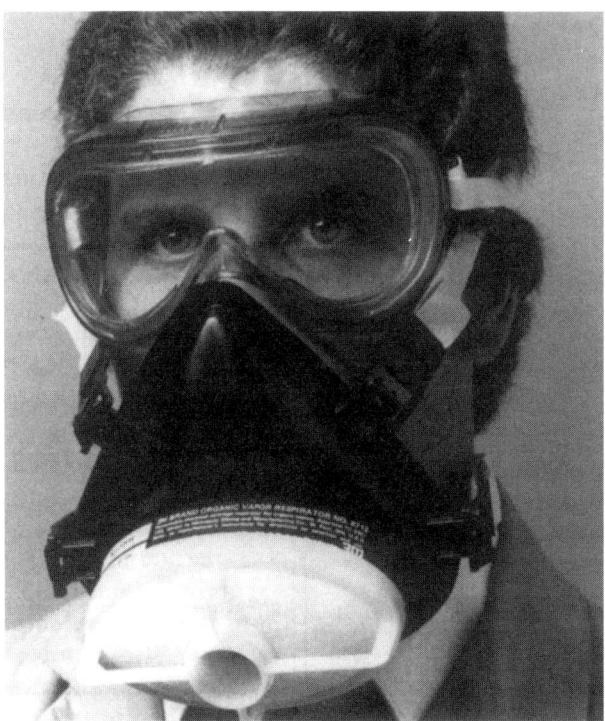

Figure 2.12   Combination particulate/vapor respirator (courtesy of 3M Canada).

using chemical-cartridge respirators when selecting the proper respiratory protection:

- They do not supply oxygen and thus cannot be worn in oxygen-deficient atmospheres.
- These respirators are designed for protection against specific gases or vapors. The wearer must read the label carefully and understand what the cartridge is designed for. A cartridge designed for acid gases may not provide adequate protection against organic vapors.
- These respirators can be used for protection against contaminants with good warning properties (i.e., smell, taste, irritation). The respirator works at 100% efficiency until the capacity of the sorbent is reached. Then the contaminant passes through the sorbent material and is inhaled by the wearer. Thus there must be a warning system that alerts the wearer that the chemical cartridge is used up, and that the wearer can be exposed to the hazardous material.
- They must not be used in atmospheres that are immediately dangerous to life or health, except when an escape is necessary.
- These respirators must be protected from the atmosphere while in storage because they tend to pick up water vapor from the air. This property may reduce or increase the effectiveness of the sorbent. Also, the sorbent will continue to pick up the contaminant even if it is left in the work environment and is not being worn. This will reduce the useful life of the respirator.

Gas Masks

The third class of air-purifying respirators is the gas mask. Gas masks are designed to protect against slightly higher concentrations of organic vapors or gases, alkaline gases, acid gases, pesticides, paint vapors and mists, radioactive particulates, dust, mists, fumes, and certain combinations of these materials. Nearly all gas masks have a full facepiece (Figure 2.13).

The basic difference between a full-facepiece chemical-cartridge respirator and a gas mask is the volume of sorbent contained within. The volume of sorbent contained in a canister is 2 to 10 times that contained in a cartridge.

There are essentially two versions of gas masks, the first one being the front- or back-mounted gas mask. With this version, a super-size or industrial-size canister is generally fastened to the user's body, either front or back, and a breathing tube connects the canister to the facepiece inlet. The volume of sorbent contained in the canister ranges from 1000–2000 cm^3. The inhalation valve is also contained in the canister and not in the facepiece, as in chemical-cartridge respirators.

The super-size canister is somewhat larger than the industrial-size canister and has approximately twice the useful time of the industrial-size canister. Both canister sizes have various types that are specific for the gas and/or vapor contaminant. They are also labeled with an expiration date and may have

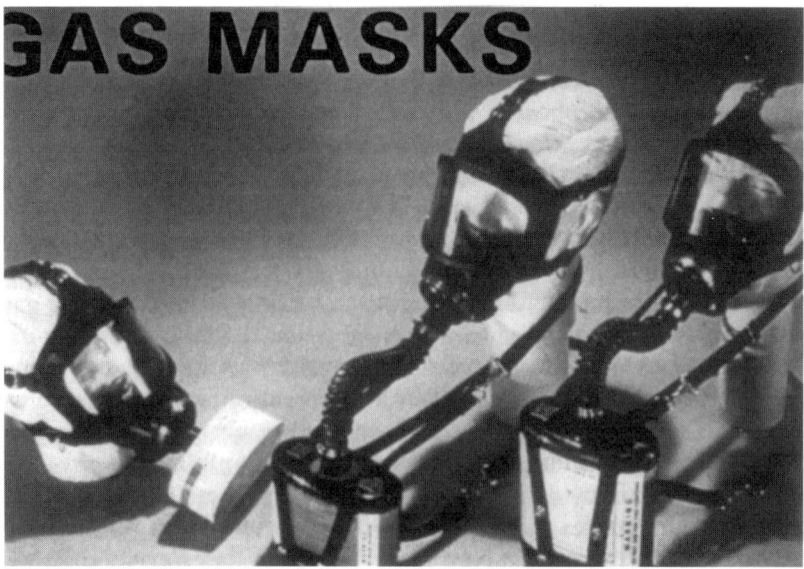

Figure 2.13   Gas masks.

a window indicator to assist in evaluating the remaining service-life. The canisters are also color-coded per the American National Standard Institute (ANSI) standard K.13.1. Again, this color-code should not be relied on solely. The user should always read the label and other pertinent information supplied with the equipment.

The second version of gas mask is the chin-style. Chin-style gas masks typically have a medium-sized canister rigidly attached to a full facepiece. The volume of sorbent contained in the canister normally ranges from 250–500 cm^3. The useful lifetime is less than that of a front- or back-mounted canister but greater than that of chemical cartridges because of the smaller sorbent volume.

Canister-type gas masks are designed for use in an atmosphere with known gas concentrations not exceeding 2% by volume (20,000 parts per million [ppm]) or as indicated on the canister label. They are generally available for protection against organic vapors, acids, gases, or ammonia. Similar to the previous two types of air-purifying respirators, gas masks do not provide protection against oxygen deficiency.

Powered Air-Purifying Respirators

The powered air-purifying respirator uses a blower to pass contaminated air through a filtering medium (Figure 2.14). It may be used for particulate, vapors, or gases, or a combination of these. The filter medium may be located in a belt-

Figure 2.14   Powered air-purifying respirator (courtesy of 3M Canada).

mount holder or in the headpiece. Air is passed through the filter by use of a blower and delivered to a facepiece or head cover (Figure 2.15). This type of air-purifying respirator has a distinct advantage because it puts the facepiece or headpiece under positive pressure with respect to the outside contaminated atmosphere, so that any leakage is outward from the facepiece. The limitations of these respirators are as follows:

- They do not supply oxygen; therefore, they cannot be worn in oxygen-deficient atmospheres.
- The type and degree of protection depends on the air-purifying element, whose protection level and useful service-time depend, in turn, on its material, size, and shape, and on the nature and concentration of the contaminant.

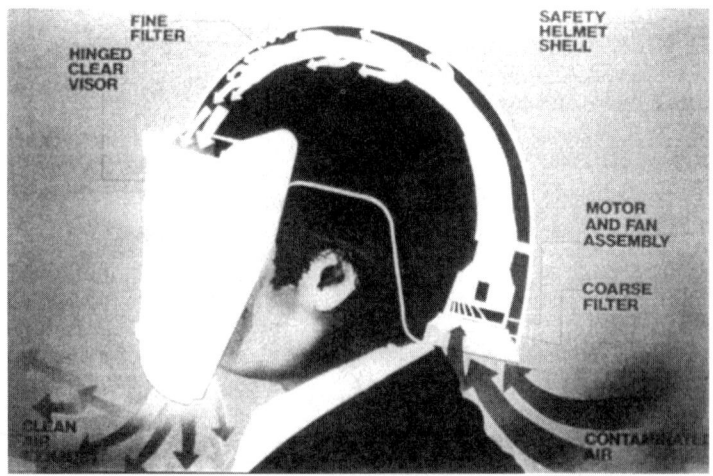

Figure 2.15   Operation of powered air-purifying respirator.

ATMOSPHERE-SUPPLYING RESPIRATORS

The atmosphere-supplying class of respirators (Figures 2.16 through 2.19) differs from the air-purifying class in that air is provided from a source independent of the surrounding atmosphere instead of purifying the atmosphere. Thus, the user is breathing within a system, which admits no outside air. There are essentially three kinds of atmosphere-supplying respirators:

1. Self-contained breathing apparatus (SCBA), where the user carries a supply of respirable air (Figures 2.16 and 2.19).
2. Supplied air-line, where the user is supplied with respirable air through a hose (Figures 2.17 and 2.18).
3. Combination SCBA and supplied air-line.

In all cases, the air supply must conform to, or meet, the requirements of the Compressed Gas Association Specification for Type 1, Class D gaseous air. This means that the carbon monoxide level must not exceed 20 ppm; carbon dioxide must not exceed 1000 ppm; and condensed hydrocarbons must not exceed $5\,mg/m^3$.

There are many different types of SCBA. These different types may be differentiated in two ways: (1) by the method by which air is supplied and (2) by the way in which the air supply is regulated.

There are essentially two ways in which air can be supplied; the first way is the rebreathing type or closed-circuit SCBA. In this type of system, the air is rebreathed after the exhaled carbon dioxide has been removed and the oxygen content restored by a compressed oxygen source or an oxygen-generating

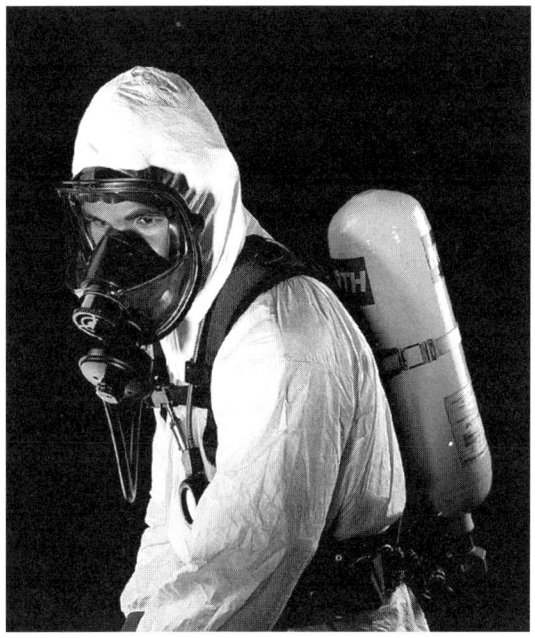

Figure 2.16 Self-contained breathing apparatus (courtesy of North Safety Products).

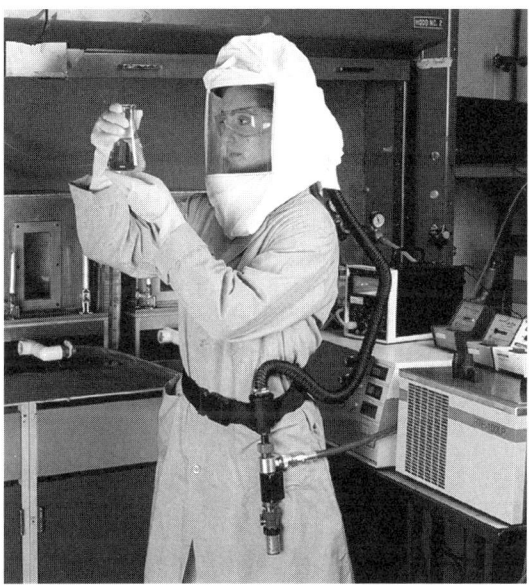

Figure 2.17 Supplied air-line respirator (courtesy of North Safety Products).

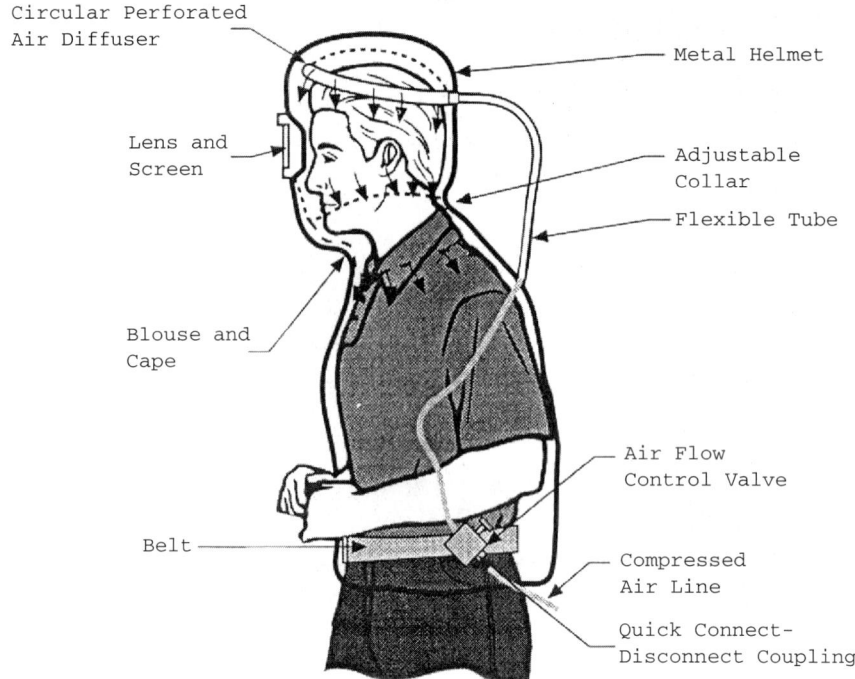

Figure 2.18 Continuous-flow air-line respirator (commonly used for abrasive blasting). (Reprinted with permission from *Respirator Protection Handbook*, Lewis Publishers, CRC Press, Boca Raton, Florida.)

substance. These devices are designed primarily for one- to four-hour use in oxygen-deficient atmospheres. Once initiated, they normally cannot be turned off. In one type of closed-circuit device, the exhaled air passes through a granular solid absorbent that removes the carbon dioxide, thus reducing the flow back into the breathing bag. The bag collapses, thus opening the admission valve to admit more pure oxygen that reinflates the bag. The oxygen, again, is supplied from a compressed oxygen or an oxygen-generating substance, such as potassium superoxide.

Another type of closed-circuit system is the oxygen-generating SCBA. This system functions differently from the compressed oxygen cylinder type because there are no mechanical operating components. This unit works on the lung-governing principle, wherein the wearers of the apparatus, by their own breathing, regulate the oxygen supply automatically and provide the exact amount required. The oxygen-generating solid is usually potassium superoxide, KO_2. The H_2O and CO_2 in the exhaled breath react with the KO_2 to release O_2. This device is front-mounted and consists of a hermetically sealed canister containing the chemical, a housing into which the canister is inserted, a breathing bag, breathing tubes, facepiece, and harness assembly. The second method in which air can be supplied is via an air cylinder, which is

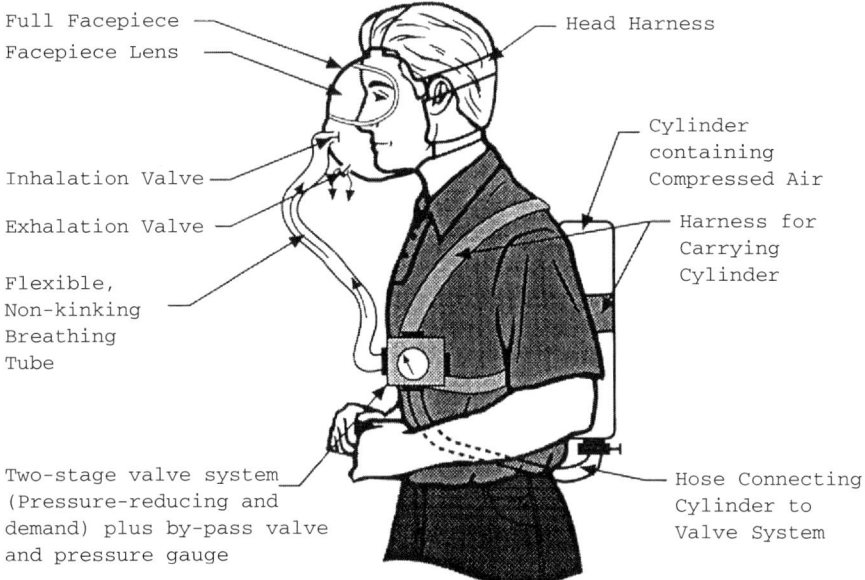

Full Facepiece
Facepiece Lens

Head Harness

Cylinder
containing
Compressed Air

Inhalation Valve

Exhalation Valve

Harness for
Carrying
Cylinder

Flexible,
Non-kinking
Breathing
Tube

Two-stage valve system
(Pressure-reducing and
demand) plus by-pass valve
and pressure gauge

Hose Connecting
Cylinder to
Valve System

Figure 2.19 Open-circuit self-contained breathing apparatus. (Reprinted with permission from *Respirator Protection Handbook*, Lewis Publishers, CRC Press, Boca Raton, Florida.)

common in the open-circuit SCBA. An open-circuit SCBA exhausts the exhaled air to the atmosphere instead of recirculating it. A tank of high-pressure compressed air, normally carried on the back, supplies air to a two-stage regulator that reduces the pressure for delivery to the facepiece. The service-life of the open-circuit SCBA is normally shorter than a closed-circuit system because it has to provide the total breathing requirements and not just the oxygen requirements for recirculation of air. Most open-circuit devices have a service-life of 30 minutes, and 15 minutes or less for special circumstances, with smaller air cylinders.

As previously mentioned, in addition to the method by which air is supplied, many different types of SCBA can also be differentiated by the way in which the air supply is regulated. Air is regulated to the facepiece by two different modes; the first being demand. In the demand-type regulator, air at approximately 200 pounds per square inch (psi) is supplied to the admission valve via a two-stage regulator that reduces the pressure to approximately 50–100 psi. The admission valve is activated upon inhalation, which creates negative pressure in the facepiece. This in turn contracts a diaphragm, causing the admission valve to open and allow air into the facepiece. The admission valve remains closed as long as positive pressure is maintained in the facepiece. In other words, air flows into the facepiece only on "demand" by the wearer, hence the name.

The second type of air regulation modes is called pressure-demand. A pressure-demand regulator is similar to the demand-type except for a spring between the diaphragm and the outside case of the regulator. This spring tends to hold the admission valve slightly open, theoretically allowing continual airflow into the facepiece. This would be true except that all pressure-demand devices have a special exhalation valve that maintains about 1.5–3 inches of water positive-back pressure in the facepiece, and opens only when the pressure exceeds that value. This combination of modified regulator and special exhalation valve maintains positive pressure in the facepiece at all times, and the regulator still supplies additional air on demand. The designed air flowrate for demand and pressure-demand regulators is approximately 350–400 liters per minute (lpm). A facepiece whose exhalation valve is designed for demand operation cannot be used with a pressure-demand regulator because air will flow continuously and quickly exhaust the air supply. Conversely, some open-circuit SCBAs are designed with the capability of switching from demand to pressure-demand.

All SCBAs may be used in oxygen-deficient atmospheres; however, only full-facepiece positive-pressure units (i.e., pressure-demand) should be used in immediately dangerous to life or health (IDLH) environments. Half-mask pressure-demand SCBA units are believed to provide no more protection against toxic substances than most air-purifying devices. All SCBA equipment must also have a functioning remaining service-life indicator or warning device that signals when only 20–25% of service-time remains.

The second kind of atmosphere-supplying respirator is the supplied air-line. Supplied air-line respirators use a central source of breathing air that is delivered to the wearer through an air-supply line or hose (Figures 2.17 and 2.18). The hose length varies from 25–300 feet, and the maximum-inlet operating pressure must not exceed 125 per square inch gauge (psig), at the point where the hose attaches to the air supply.

The air-line respirators may consist of a half-mask, full facepiece, hood, or helmet, or as a complete suit in certain applications. Full-suit air-line respirators are used against substances that irritate, corrode, or penetrate through the skin, and they also provide breathing air. Designs for abrasive blasting are also available, which protect against impact and abrasion from rebounding material.

Air-line respirators are available in demand, pressure-demand, and continuous-flow configurations. A demand or pressure-demand air-line respirator is similar to a demand or pressure-demand open-circuit SCBA, except that the air is supplied through a small-diameter hose from a stationary source of compressed air rather than from a portable high-pressure hose. Regulators for air-line respirators also have only single-stage reduction as opposed to two-stage reduction because the air pressure is limited to 125 psi.

Continuous-flow air-line respirators maintain airflow at all times, rather than only on demand. In place of a demand or pressure-demand regulator, an airflow control valve or orifice partially controls the airflow. To be NIOSH approved, the respirator must deliver at least 115 liters (4 cubic feet) of air per minute for tight-fitting facepieces and 170 liters (6 cubic feet) of air per minute for loose-fitting facepieces, hoods, or helmets.

The stationary source of compressed air could be from a compressed-air cylinder; a compressor; or purified, compressed plant air. Caution should be exercised to ensure that the breathing air is free of unwanted contaminants and meets the requirements of Compressed Gas Association Specification for Type 1, Class D gaseous air.

Supplied air-line respirators provide a high degree of protection, but their use is limited to non-IDLH atmospheres because the wearer is totally dependent on the integrity of the air-supply hose and must be able to escape from the contaminated area without endangering his life. The last kind of an atmosphere-supplying respirator is the combination SCBA and supplied air-line. This type is essentially the same as the air-line respirator, with an added compressed-air cylinder that may be carried on one's back or at one's side in a sling. Combination units may consist of a half-mask or full facepiece and operate in the demand or pressure-demand mode.

Combination units are generally used for emergency entry into and escape from IDLH atmospheres. The self-contained part of the device is used only when the air-line part fails and the wearer must escape, or when it may be necessary to disconnect the airline temporarily while changing locations. A combination air-line and SCBA may also be used for emergency entry into a hazardous atmosphere to connect to an air-line, if the SCBA is classified for service of 15 minutes or longer and not more than 20% of the air supply's rated capacity is used during entry.

CONCLUSION

In summary, the basic purpose of a respirator is to protect the respiratory system from harmful airborne contaminants. It does this by removing—through mechanical filtration, chemical reaction, adsorption, or absorption—the contaminant from the air, or by supplying an independent source of breathable air. Selection of the proper type of respirator provides workers with the necessary protection they require. There are limitations to all types of respirators that must be understood. This chapter has summarized the different types of respirators available and their corresponding limitations. More detailed information on specific respirators or types of respirators can be obtained from respirator manufacturers. The listing for Particulate Respirators Certified under 42CFR84 can be found at the Website: **www.cdc.gov/niosh/p8intro.html.**

REFERENCES

1. Revoir, William H., and Bien Ching-Tsen. *Respirator Protection Handbook*. Boca Raton, FL: Lewis Publishers, Division of CRC Press, 1997.
2. National Institute of Occupational Safety and Health (NIOSH) Guide to Industrial Respiratory Protection. DHHS (NIOSH) Publication No. 87-116. Cincinnati, OH: NIOSH, 1987.
3. Code of Federal Regulations, "Respiratory Protection" Title 29, Part 1910.134. pp. 412–437, 1998.
4. Rajhans, G.S, and D.S.L. Blackwell. *Practical Guide to Respirator Usage in Industry*. Boston: Butterworth–Heinemann, 1985.

3

Criteria for Selection and Fitting

After presenting detailed information on respirator types and limitations in Chapter 2, we begin this chapter with the regulatory requirements that govern the use of respiratory protection in the workplace. In Chapter 7, a respiratory protection program is discussed, while this chapter and Chapters 4, 5, and 6 expand on the program elements. The key components of the respiratory protection program have come about because of regulatory requirements, so it is worthwhile for readers to be familiar with these regulatory requirements.

The various regulatory authorities in industrialized nations appear to have accepted the fact that very low acceptable levels of exposure to airborne contaminants cannot be achieved with engineering controls alone, and hence, must be augmented by respiratory protective equipment (RPE). This realization has prompted these countries to establish various criteria and guidelines for testing and approving respirators. Some of these guidelines are minimum requirements for employers in order to provide an adequate respiratory protection program for employees. Some of these minimum requirements may include the following:

- Written procedures for selection and use of respirators, fitting, training, and cleaning
- Preuse and annual medical evaluation to ensure that the employee is physically able to wear and work with the equipment
- Selection and use of respirators on the basis of the hazard(s) involved
- Training program designed to train workers in the correct method of wearing and using a respirator
- Procedure for use of respirators in routine work and in emergency situations
- Regular, periodic inspecting, cleaning, repairing, and disinfecting of used respirators
- Procedure to ensure adequate quality of breathing air for atmosphere-supplying respirators
- Training of employees in the respiratory hazards they are potentially exposed to while performing routine tasks and in emergency situations
- Procedures for evaluating the effectiveness of the respiratory protection program

The requirements also state that approved or accepted respirators should be used when they are available.

FACTORS THAT CAN INFLUENCE RESPIRATOR SELECTION

The following factors can influence respirator selection:

1. *Physical configuration of the workplace.* The feasibility of using a particular type of respirator should be examined based on workplace conditions. For example, it may not be practical to use self-contained breathing apparatuses (SCBAs) in tightly constrained spaces, even though they might be an acceptable choice otherwise. Similarly, use of air-line respirators may not be practical in areas with obstructions or moving machinery that can snag air-supply hoses.
2. *Worker medical condition.* Wearing respiratory protection poses a physical burden on the wearer. Therefore, workers with breathing problems may not be able to use negative-pressure respirators.
3. *Worker comfort.* Worker preferences such as breathing ease, skin comfort, in-mask temperature and humidity, weight, and convenience should be considered during respirator selection.

In theory, a regulatory authority may accept a respirator evaluated by any competent authority; however, in North America only, three organizations have been historically recognized as competent: the National Institute for Occupational Safety and Health (NIOSH), the Mine Safety and Health Administration (MSHA), and the British Standards Institute (BSI).

HISTORY OF APPROVAL OF RESPIRATORS IN UNITED STATES

In 1919, the Bureau of Mines (BOM) brought out the first approval schedule. This included:

SCHEDULE: 13. Self-Contained Breathing Apparatus (SCBA)
             14. Gas Mask (full-face mask)
             19. Air-Line Respirators
             21. Dust, Fume, and Mist Respirators (D.F.M.)
             23. New Emergency Organic Vapor Respirators, O.V.

As the schedules were revised over the years, letters were included with the numbers (e.g., 13A, 13B). There were two problems with this scheduling system:

1. There were no provisions for decertification (e.g., a CHEMOX oxygen breathing apparatus purchased under Schedule 13, in 1919, would still be approved today).
2. More than half of the respirators were being used above ground in nonmining operations.

NIOSH and MSHA started the present respirator-approval program in 1971 when they assumed BOM responsibilities. All testing for approvals is performed by NIOSH, and the results are reviewed by NIOSH and MSHA, which are the approval agencies. The Williams and Steiger Occupational Safety and Health Act (OSHA) was enacted in 1970. It sets forth specific legal requirements for selection, use, and maintenance of respirators, and gives guidelines for establishing a respirator program to meet these requirements. These requirements are outlined in Title 30, Code of Federal Regulation, Part II [30 CFR part II], commonly known as "Part II".[1] Schedules and terminology have changed as follows:

Part II, Subpart H—Self-Contained Breathing Apparatus (Old Schedule 13F)
Subpart I—Gas Masks (Schedule 14G)
Subpart J   Supplied-Air Respirators (Schedule 19C)
Subpart K—Dust, Fume, and Mist Respirators (Schedule 21C)
Subpart L—Chemical-Cartridge Respirators (Schedule 23C)
Subpart M—Pesticide Respirators (Schedule 23C)

In the transition period, respirators with BOM approval could be used until March 30, 1974; however, grandfather clauses and extensions allowed some BOM respirators to be used legally (e.g., gas masks under Schedule 14F, self-contained breathing apparatus under Schedule 13-13F [good until March 31, 1979]).

In 1987, NIOSH published its *Guide to Industrial Respiratory Protection*,[2] which serves as an update of the 1976 NIOSH publication *A Guide to Industrial Respiratory Protection*.[3] On July 10, 1995, 30CFRII was replaced by 42CFR84[4] as an active regulation. All nonpowered, air-purifying, particulate-filter respirators approved under Part 84 must meet the new performance standard; however, a new sequence of approval numbers (TC-84A-XXXX) is used for nonpowered particulate respirators certified under Part 84. All other respirator types will continue to use the sequence of approval numbers previously used for Part II because the requirements for these other types have not changed. For example, the number series TC-13F-XXXX indicates an SCBA that is certified under the provisions of either the old Part II or the new Part 84. Similarly, powered air-purifying respirators for particulates that are certified under the new Part 84 will continue to be numbered with the sequence TC-21C-XXXX (as they were numbered under Part II) because the certification requirements have not yet changed. All particulate respirators approved under Part 84 will have a certification label bearing the NIOSH and the Department of Health and Human Services (DHHS) emblems, whereas those approved under Part II have the emblems of NIOSH and MSHA. This allows the user to distinguish particulate respirators certified before July 10, 1995, under Part II, from particulate respirators certified after that date under Part 84.

The revised testing requirements for particulate filters are much more demanding than the old Part II tests, and they provide much better evidence of the filter's ability to remove airborne particles. The new requirements are consistent with 20 years of advances in respiratory protection technology.

Generally, it is difficult to determine if a respirator (in total) is approved unless a person is experienced. Simply put, there must be absolutely no change to the respirator (e.g., missing parts, incorrect parts) that has passed the approval schedule, or else that respirator is not approved. The respirator(s) in question must be checked against manufacturers' literature and the *NIOSH Certified Equipment List* (**www.cdc.gov/niosh/celpamp.html**), which is updated every year.[5]

In addition to OSHA requirements for testing and approval, there is also the joint NIOSH/OSHA standard completion program, which forms the basis for Respirator Decision Logic.[6] The purpose of decision logic is to provide the necessary criteria to support the selection of proper respiratory protection. It is a step-by-step procedure that eliminates inappropriate respirators and eventually leads to the correct choice of equipment. Protection factors are the criteria used to determine to what maximum concentrations the selected respirator may be worn. Reference to this NIOSH/OSHA Respiratory Decision Logic[6] and the Title 42 CFR Part 84[4] should be made when RPE is being selected for the first time.

Thus, there are specific OSHA requirements for respirator approvals, including fit testing, respirator usage, care and maintenance, training and fitting, program administration, and surveillance and program evaluation. OSHA has occasionally changed its requirements for qualitative and quantitative fit testing of respirators used to protect against a specific toxic substance. Similarly, NIOSH constantly issues notices and warnings concerning respirators. We strongly recommend that readers review the Websites given in the Appendix.

SUMMARY OF BRITISH STANDARDS INSTITUTION REQUIREMENTS

British standards, like other respiratory standards, are designed to assist in selecting respirators to protect against atmospheric contaminants, such as dust and gas, and against oxygen deficiency. British standards have been revised several times under the following designations:

- **BS EN 372:1992**—Specification for SX gas filters and combined filters against specific named compounds used in respiratory protective equipment[7]
- **BS EN 371:1992**—Specification for AX gas filters and combined filters against low boiling organic compounds used in respiratory protection equipment[8]

- **BS EN 143:2000**—Respiratory protective devices. Particle filters. Requirements, testing, marking[9]
- **BS EN 136:1998**—Respiratory protective devices. Full face masks. Requirements, testing, marking[10]
- **BS EN 145:1998**—Respiratory protective devices. Self-contained closed-circuit breathing apparatus compressed oxygen or compressed oxygen-nitrogen type. Requirements, testing, marking[11]
- **BS EN 137:1993**—Specification for respiratory protective devices: self-contained open-circuit compressed air breathing apparatus[12]
- **BS EN 270:1995**—Respiratory protective devices. Compressed air-line breathing apparatus incorporating a hood. Requirements, testing, marking[13]
- **BS EN 400:1993**—Respiratory protective devices for self-rescue. Self-contained closed-circuit breathing apparatus. Compressed oxygen escape apparatus. Requirements, testing, marking[14]
- **BS EN 12941:1999**—Respiratory protective devices. Powered filtering devices incorporating a helmet or a hood. Requirements, testing, marking[15]
- **BS EN 149:2001**—Respiratory protective devices. Filtering half masks to protect against particles. Requirements, testing, marking.[16]

The BSI warns users of the standards to refer to the latest edition of these standards, including amendments (Website: http://bsonline.techindex.co.uk). In addition to the preceding standards, the BSI also published recommendations and specifications for fit test, selection, use, care, and maintenance of RPE. For example, BS 4275:1997, Guide to implementing an effective respiratory protective device programme for establishing a respiratory program.[17]

BS EN 149:2001 provides detailed specifications for filtering-facepiece dust respirators.[17] BS 4275 gives recommendations for the selective use and maintenance of respiratory protective equipment (RPE),[17] and BS 4400[18] provides details for sodium chloride tests for respirator filters. The BSI provides certification for a product created to ensure personal safety and quotes the following in their specification:

A license to use the Kitemark or Safety mark on or in relation to a product will be granted to any manufacturer or producer who demonstrates that he/she can and will be able consistently to make that product to the requirements specified in the British Standard. His/her capability of doing so is initially assessed by inspection of his/her production process, quality control organization, and test facilities, and by independent testing of a sample of the product against all the criteria of the relevant standard. The licensee is required to accept and to operate in accordance with a BSI scheme of supervision and control, which identifies the minimum level of quality control to be exercised during manufacture and the tests to be carried out on the completed product. BSI carries out unannounced inspection visits to the manufacturer's works and audit-testing of the product, and may withdraw the license for any failure of the manufacturer to comply with the relevant standard or the requirements of the scheme of supervision and control. The presence of the mark on or in relation to a product is an assurance that the goods have been produced under a system of supervision, control and testing, operated during manufacture and including periodical inspection of the manufacturer's works, in accordance with the certification mark scheme of BSI.

No RPE can provide optimum performance if it leaks, and one of the main sources of leakage is poor fit of the mask on the face. As a result of both field and laboratory studies into respirator performance, guidance on regulations (such as the Control of Asbestos at Work regulations) and general HSE (Health and Safety Executive, UK) guidance[19] on the use of RPE (HSG53) now strongly recommend that fit testing be included as an integral part of an RPE program. The best time to conduct fit testing is at the initial selection stage, when individual users should be given a choice of suitable models of RPE.

Other countries, such as Japan, Germany, and Italy, have standards for RPE that outline specification and performance requirements. Workers must familiarize themselves with the regulatory requirements governing the use of RPE in the workplace. We recommend that you contact your local occupational health and safety agency for specific details on regulatory requirements for the use of respirators.

CRITERIA FOR SELECTION

Under normal circumstances, few manufacturing operations or maintenance procedures require the use of a respirator. Ordinarily, it is more practical and less costly to use process-engineering controls, good work practices, and less toxic materials to maintain employee exposures at acceptable levels. All Occupational Health and Safety Acts in North America require that air-contamination levels be reduced to the lowest feasible level by engineering controls and process design. When this is not possible, or while such controls are being implemented, the appropriate respirators may be used to protect employees. See the Respirator-Use Requirements Flow Chart (Figure 3.1).

General Guidelines

These guidelines were formulated according to OSHA Standards—29CFR 1910 134 Respiratory Protection.[20]

Jobs that may require the use of RPE include, but are not limited to, the following:

- Grinding, cutting, or otherwise machining of solid materials that may liberate significant quantities of uncontrolled dust
- Manufacturing operations that require entry into oxygen-deficient or potentially oxygen-deficient atmospheres
- Maintenance in areas processing, handling, storing, or disposing of potentially toxic materials
- Under emergency conditions (e.g., spills, leaks) for escape, rescue, repairs, or shutdown

Respirator-Use Requirements Flowchart 29 CFR 1910.134 (c)

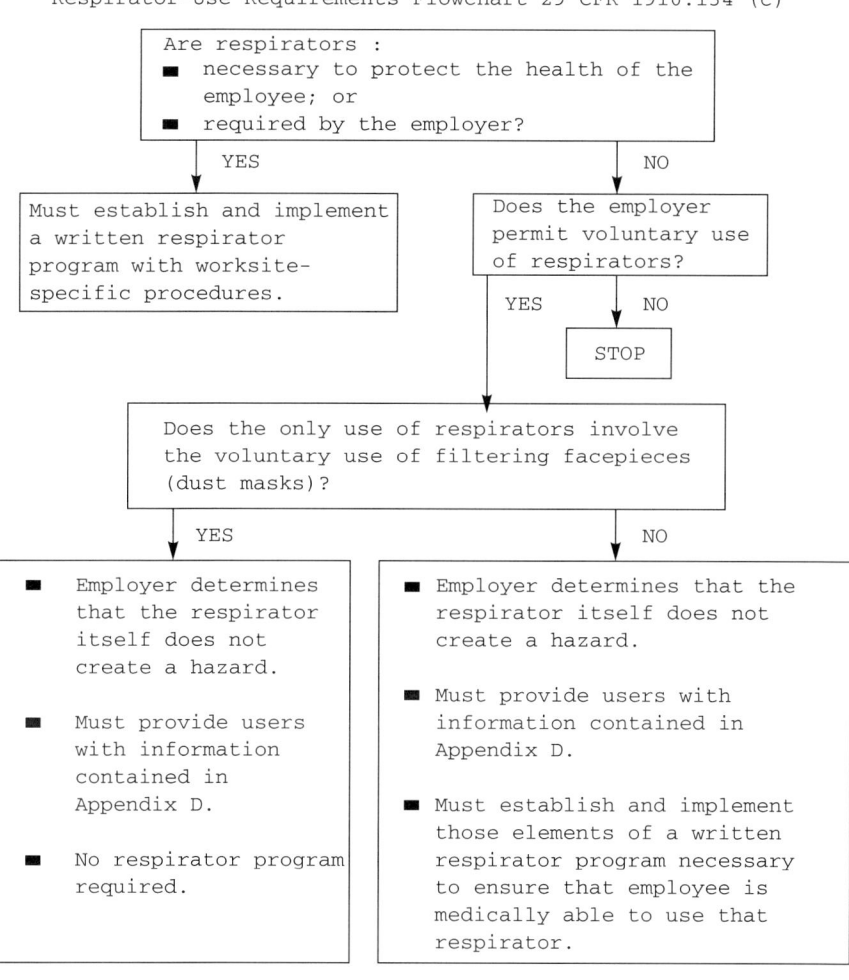

Figure 3.1   Respirator-use requirements flow chart.[21]

When respirators are used, it is essential that a respiratory protection program be established. Proper respirator selection can be one of the most difficult jobs for the health and safety professional in a respiratory protection program. There are several different types of respiratory hazards and certain limitations to differentiate types of respirators. It is essential that the proper respirator be selected. Proper respirator selection is the choice of a device that (1) fully protects the worker from the hazards to which he may be exposed, (2) provides the greatest reliability possible, and (3) permits the worker to perform the job with the least amount of discomfort and fatigue.

Consider the following steps when selecting respirators:

1. Evaluate respiratory hazard(s) in the workplace; identify relevant workplace and user factors.
2. Provide an appropriate respirator based on the respiratory hazard(s) to which the worker is exposed and workplace and user factors that affect respirator performance and reliability.
3. Select a NIOSH-certified respirator and use it in compliance with the conditions of its certification.
4. Evaluate the respiratory hazard(s) in the workplace, including a reasonable estimate of employee exposures to respiratory hazard(s) and an identification of the contaminant's chemical state and physical form. Where the employee exposure cannot be identified or reasonably estimated, the employee atmosphere should be considered immediately dangerous to life or health (IDLH).
5. Select respirators from a sufficient number of respirator models and sizes so that the respirator is acceptable to, and correctly fits, the user.
6. Obtain a written recommendation regarding the employee's ability to use the respirator from a physician or other licensed health care professional.

Selection of the proper type of respirator for any given situation should require consideration of the following facts:

*   The nature of the hazard
*   The characteristics of the hazardous operation or process
*   The location of the hazardous area with respect to a safe area having respirable air
*   The period for which respiratory protection may be required
*   The activity of workers in the hazardous area
*   The physical characteristics, functional capabilities, and limitations of respirators of various types
*   The respirator fit factor and respirator fit

Nature of Hazard

The following conditions should be considered in the selection procedure:

*   Type of hazard
    *   Oxygen deficiency
    *   Contaminant
*   Physical properties (gas, vapor, dust, mist, fume)
*   Chemical properties
*   Physiological effects on the body
*   Actual concentration (average, peak) of the toxic material. Recognition and evaluation of the respiratory hazard is an essential part of selecting a

respirator. Initial monitoring of the respiratory hazard must be carried out to obtain data needed to select proper respiratory protection. This would include identification of the type of respiratory hazard, the airborne concentration, and the nature of the contaminant.

- Established permissible time-weighted average or peak concentration of the toxic material
- Whether the concentration of toxic material presents a hazard that is IDLH. Two facts are considered when establishing IDLH concentrations.
  - The worker must be able to escape without losing his life or suffering permanent health damage within 30 minutes.
  - The worker must be able to escape without severe eye or respiratory irritation or other reactions that could inhibit escape.

If the concentration is above the IDLH, only a highly reliable breathing apparatus, such as positive-pressure demand is allowed.

- Warning properties, such as odor, eye irritation, and respiratory irritation, that rely on human senses for detection are not foolproof; however, they do provide some indication to the wearer that the service-life of the cartridge or canister is reaching the end, the facepiece is not fitted properly, or there is some other respirator malfunction. Warning properties may be assumed to be adequate when the effects of odor, taste, or irritation from the substance can be detected and are persistent at concentrations or at below the permissible exposure level. If the odor or irritating threshold of the material is many times greater than the permissable-exposure limit, the substance is considered to have poor warning properties, and air-supplied respirators must be specified. Table 2.3, Chapter 2, contains pertinent information on odor threshold limit values for several contaminants.
- Skin absorption of the contaminant must be considered. Skin protection must be provided when information is available indicating possible systematic injury or death resulting from absorbance of a gas or vapor through the skin. This may take the form of impervious coverings or supplied-air suits.
- Eye irritation must also be considered. Eye protection may be required in the case of certain concentrations of acid gases and organic vapors. Full-face respirators may be needed to aid in protecting the eyes and the respiratory tract.

Characteristics of Hazardous Operations or Process

In selecting the proper respirator, the operation or process in which the respirator has to be worn must be closely scrutinized. Workers' activities, the characteristics of the operation or process, and work area characteristics must be considered. Materials used, including raw materials, end products, and by-products, must be known in order to select the proper respirator. Any modification in the operation should be investigated because this may change the hazard and/or the degree of the hazard, and thus a different respirator may be needed.

Location of Hazardous Area

Are there safe areas where respirable air is present near the hazardous area? This should be considered when selecting the proper respiratory equipment. Depending on the location of the hazardous area, the need to escape in case of a leak or accident and the possibility of the need for rescue may influence the type of respirator that is needed.

Respirator Usage Time

Consideration must be given to the application of the respirator. Is it routine, nonroutine, emergency, escape, or rescue use? The length of time the respirator has to be worn will influence selection. Lightweight, comfortable respirators should be selected for routine wearing over long periods during the work shift.

Worker Activity in Hazardous Area

Worker activity and location in the hazardous operation should be examined. The selection of a proper respirator will be influenced by whether the work load is light, medium, or heavy and whether the worker is in the area continuously or irregularly.

Respirator Characteristics, Capabilities, and Limitations

The characteristics, capabilities, and limitations of different types of respirators (detailed in Chapter 2) must be considered in the selection procedure. The preceding six conditions must be considered when selecting the proper respiratory equipment to ensure the protection of the workers in the hazardous work area. Figure 3.2 summarizes the steps necessary in selecting the proper respirator. Specific respirator selection sequences are summarized in Tables 3.1 to 3.3.[22]

Respirators for IDLH Atmospheres

For protection against gases and vapors, provide an atmosphere-supplying respirator or an air-purifying respirator if the respirator is equipped with an end-of-service-life indicator (ESLI) certified by NIOSH for the contaminant; or, if there is no ESLI appropriate for conditions of the employer's workplace,

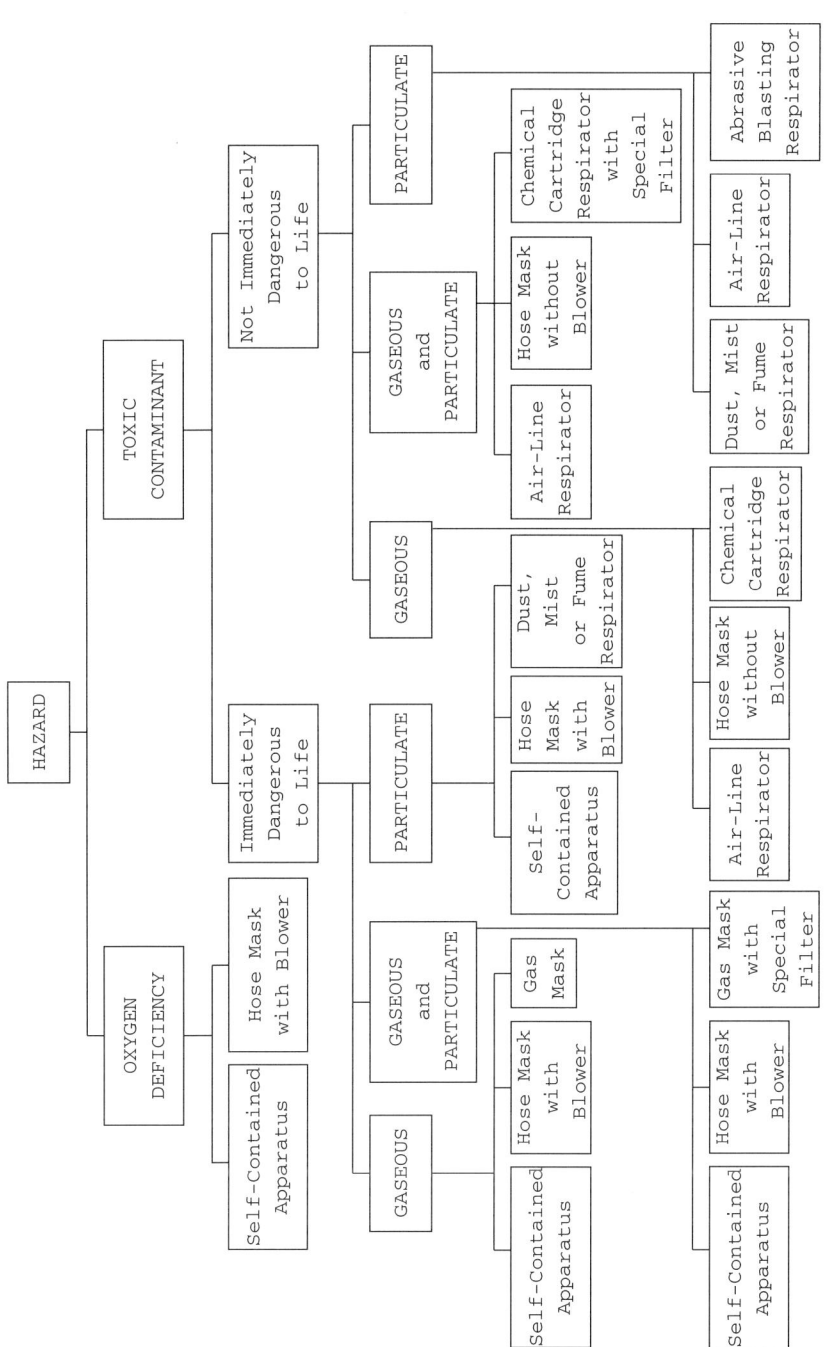

Figure 3.2 Respirator selection according to hazard.

Table 3.1   For respiratory protection against gases or vapors

APPLICATION	SELECTION SEQUENCE
Routine use	1. Consider absorption through the skin. 2. If the contaminant has poor warning properties, eliminate all air-purifying respirators. 3. If eye irritation occurs, eliminate or restrict the use of a half-mask respirator. 4. If concentration is greater than IDLH or LEL, eliminate all but positive-pressure SCBA and combination positive-pressure SCBA. 5. List suitable respirators by condition of use and type.
Entry and escape from unknown concentrations	Use positive-pressure SCBA or positive-pressure supplied-air respirator with positive-pressure SCBA.
Firefighting	Use positive-pressure SCBA.
Escape	Use gas mask or escape-SCBA.

Table 3.2   For respiratory protection against particulates

APPLICATION	SELECTION SEQUENCE
Routine use	1. Consider absorption through the skin. 2. If eye irritation occurs, eliminate or restrict use of the half-mask respirator. 3. When the maximum permissable exposure is less than $0.5 \, mg/m^3$, eliminate dust, fume, and mist (DFM) respirators unless they are fitted with high-efficiency particulate air filters. 4. If the concentration is greater than IDLH or LEL, eliminate all but positive-pressure SCBA and combination positive-pressure SCBA. 5. List suitable respirators by condition of use and type.
Entry and escape from unknown concentrations	Use positive-pressure SCBA or combination positive-pressure supplied-air respirator with positive-pressure SCBA.
Firefighting	Use positive-pressure SCBA

implement a change schedule for canisters and cartridges that ensures they are changed before the end of their service-life and describes in the respirator program the information and data relied on and the basis for the change schedule and reliance on the data. Provide a full-facepiece pressure-demand SCBA certified by NIOSH for a minimum service-life of 30 minutes or a combination full-facepiece pressure-demand supplied-air respirator with auxiliary self-contained air supply.

Table 3.3 For respiratory protection against a combination of gas or vapor and particulates

APPLICATION	SELECTION SEQUENCE
Routine use	1. Consider absorption through the skin. 2. If the contaminant has poor warning properties or inadequate sorbent efficiency, eliminate all air-purifying respirators. 3. Eliminate all respirators, except those with combination sorbent/particulate filters. 4. If eye irritation occurs, eliminate or restrict the use of the half-mask respirator. 5. When the maximum permissible exposure is less than $0.05\,mg/m^3$, eliminate all respirators except those fitted with sorbent/high-efficiency particulate air filters. 6. If the concentration is greater than IDLH or LEL, eliminate all but positive-pressure SCBA and combination positive-pressure SCBA. 7. List suitable respirators by condition of use and type.
Entry and escape from unknown concentrations	Use positive-pressure SCBA or combination positive-pressure supplied-air respirator with positive-pressure SCBA.
Firefighting	Use positive-pressure SCBA.
Escape	Gas mask or escape-SCBA.

Provide NIOSH-certified-for-escape respirators for escape from IDLH atmosphere. Consider all oxygen-deficient atmospheres to be IDLH. Unless oxygen concentration can be maintained within the ranges between 19.5% and a lower value that corresponds to an altitude-adjusted oxygen partial pressure equivalent to 16% oxygen at sea level, then any atmosphere-supplying respirator may be used.

Respirators for Non-IDLH Atmospheres

Provide a respirator that is adequate to protect the health of the employee and ensure compliance with regulatory requirements, under routine and reasonably foreseeable emergency situations. Select a respirator that is appropriate for the chemical state and physical form of the air contaminant.

Respirators for Protection against Particulates

Provide an atmosphere-supplying respirator or an air-purifying respirator equipped with a filter certified by NIOSH under 42 CFR part 84; or, for contaminants consisting primarily of particles with mass median aerodynamic diameters (MMADs) of at least 2 micrometers, an air-purifying respirator equipped with any filter certified for particulates by NIOSH.

FITTING

Considering the characteristics of the hazard involved and the capabilities and limitations of the respirator allows for the selection of a proper type of respirator, but it will not ensure that the worker is protected. Proper face-fit of the respirator is the last factor to consider in the selection procedure and is essential. Wearing a poorly fitting respirator may be more dangerous than not wearing a respirator at all. The worker may think he is protected, when in reality he is not. The following factors affect the fit of a respirator:

- Design of the respirator
- Facial features and facial hair
- Change in physical features
- Training

The term *protection factor*[23,24] has been used for many years to refer to the measure of respirator performance. This term has been misused in the past few years. The protection factor for a respirator is a fraction calculated as the amount of contaminant outside the respirator compared to that inside the respirator. This factor is the number assigned to a particular type of respirator or to an entire class of respirators, representing the degree of protection the respirator is thought to provide for most users. Protection factor determinations are not made for an individual respirator wearer. Much research has been done to assign protection factors to the various types of respirators. Table 3.4 shows assigned protection factors (APFs) given in a recent publication.[25]

Two other terms that are used to describe the respirator's protection ability are fit factor and field performance factor.

Fit factor is the measure of the sealing of a respirator to the face of the wearer, as determined by the quantitative fit test. It is the ratio of the concentration of the test atmosphere outside the respirator to the concentration inside. Fit factor is only a measure of the effectiveness of the seal between the face and the respirator. This factor is assigned to a particular individual and respirator facepiece based on individual testing; it gives some indication of the expected performance of the respirator on the wearer in the workplace.

The *field performance factor* is a measure of the actual protection provided by the respirator while it is worn in the workplace. The field performance factor is the ratio of the contaminant outside the respirator to that on the inside. Factors that affect the field performance factor include facepiece sealing, efficiency of the air-cleaning elements, training and attitude of the wearer, comfort of the respirator, and ease of putting on and adjusting the respirator.

Table 3.4 Assigned protection factors (APFs) recommended by NIOSH, ANSI standard Z88.2–1992, and OSHA[25]

CLASS OF RESPIRATOR	NIOSH	ANSI-1992	OSHA*
Half facepiece			
Any dust filter	10	10	
HEPA or 100% filter	10	10	10
Chemical cartridge	10	10	
Full facepiece			
Any dust filter	10	100	
HEPA or 100% filter	50	100	50
Chemical cartridge	50	100	
Powered air-purifying respirator			
Half facepiece			
Any particulate filter	50	50	
HEPA or 100% filter			50
Chemical cartridge	50	50	
Full facepiece	50	100	
HEPA or 100% filter chemical cartridge	50	1000	250
Helmet/Hood			
Any particulate filter	25	100	
HEPA 100% or filter	25	1000	25
Chemical cartridge	25	1000	
Loose-fitting facepiece			
Any particulate filter	25	25	
HEPA 100% or filter	25	25	25
Chemical cartridge	25	25	
Supplied air respirator, continuous flow			
Half facepiece		50	50
Full facepiece		1000	250
Hood or helmet	25	1000	25
Loose-fitting facepiece	25	25	
Supplied-air respirator, demand flow			
Half facepiece	10	10	
Full facepiece	50	100	
Hood or helmet	25		
Loose-fitting facepiece	25		
Supplied-air respirator, pressure demand or positive pressure			
Half facepiece	1000	50	1000
Full facepiece	2000	1000	1000
Supplied-air respirator, pressure demand or positive pressure full facepiece with auxiliary SCBA operated in pressure demand or positive pressure	1000		>1000
SCBA, pressure demand or positive pressure with full facepiece	10,000		>1000

*APF values are taken from the cadmium standard 1910.1027[26], but they do vary from standard to standard.

Respirator fit, simply stated, is the ability of the device to interface with the wearer to prevent the workplace atmosphere from entering the wearer's respiratory system through that interface. When the contaminated atmosphere penetrates the interface and enters the wearer's breathing zone, exposure occurs. Assessment and control must be made of the exposure resulting from an imperfect fit in order to effectively utilize respirators as a means to control airborne hazards. There is professional agreement on the need to assess and control respirator fit, but there is very little agreement on the methods to be used and control levels to be attained.

Currently, there are four classes of methods to assess respirator fit:

1.   Positive-/negative-pressure checks
2.   Qualitative fit tests
3.   Quantitative fit tests
4.   Protection-factor tests

Positive-/Negative-Pressure Checks

Pressure checks are the most commonly used method; however, they are the least precise and are applicable only to tight-fitting respiratory-inlet coverings. The positive-pressure method involves having the wearer close the outlet of the respirator (Figure 3.3). The wearer checks either for outward leakage around the sealing edge or the length of time the pressure is maintained in order to judge the adequacy of the seal.

The negative-pressure method involves the wearer closing the inlet of the respirator and then inhaling, creating a relatively high negative pressure inside the facepiece (Figure 3.4). The same criteria as those of the positive-pressure check are used by the wearer to judge the adequacy of the seal.

Although these checks detect only large leaks, they are the only method that is easily administered by the user, and thus are the most feasible method for checking fit each time the respirator is put on.

The advantages of these tests are that they are simple and quick, can be carried out by the wearer, and can be performed after each donning of the respirator during a workshift.

The limitation of these tests is that only large leaks can be detected. Care must be taken not to force the respirator against the face of the wearer, thereby creating an unrealistic sealing force with the result of low leakage.

Qualitative Fit Tests

Qualitative fit tests are often used in industry. They provide a valid assessment of fit for negative-pressure respirators used for protection against toxic contaminants.

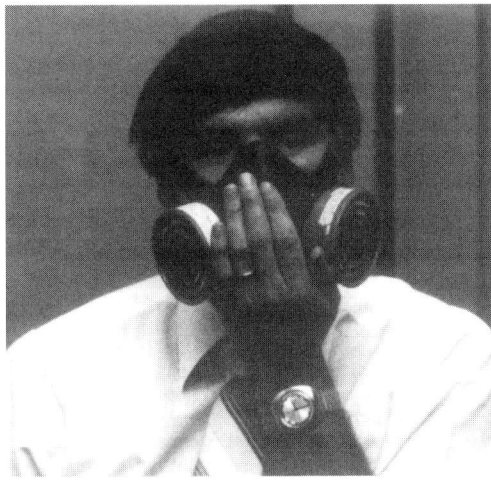

Figure 3.3 Positive-pressure face-fit test on a dual-cartridge respirator.

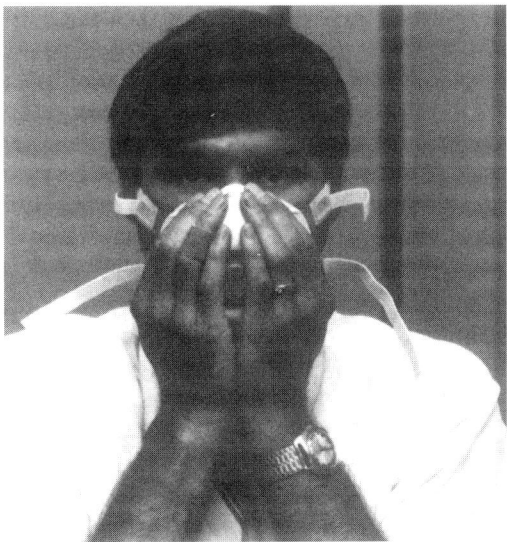

Figure 3.4 Negative-pressure face-fit on a disposable respirator.

There are three qualitative methods, all of which generate a test atmosphere around the respirator, attempting to concentrate it around the sealing area. If a significant leak exists, the wearer senses the presence of the test agent; if the test agent is not detected, the respirator is judged to have an adequate fit. Detailed procedures for qualitative fit testing can be found in OSHA standard 29 CFR 1910.134.[21]

Irritant Smoke Test

This method is commonly used with air-supplied respirators or air-purifying respirators equipped with high-efficiency air-purifying elements. The "smoke" is generated by aspiring moist room air through a stannic chloride ventilation test tube. The smoke consists of HCl on a very small oxide particle that is directed around the sealing surface of the respirator. If a leak is present, the wearer detects it by reacting to the irritation, usually with an involuntary cough. The person to be tested must demonstrate his or her ability to detect a weak concentration of the irritant smoke (see Figure 3.5).

The main advantage of this method is that it usually evokes an involuntary response in the form of a cough or sneeze. The likelihood of this giving a false indication of proper fit is reduced. The main limitation with irritant smoke is that the test must be performed with caution because the aerosol is highly irritating to the eyes, skin, and mucous membrane.

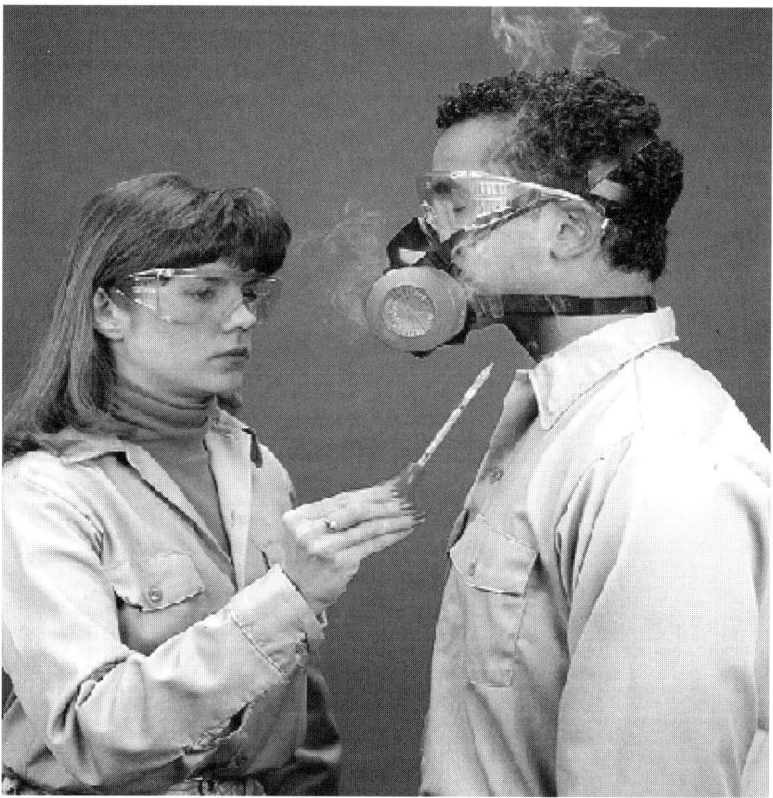

Figure 3.5   Irritant smoke fit test (courtesy of North Safety Products).

Isoamyl Acetate or Banana Oil Method

Isoamyl acetate is a liquid that smells strongly of bananas. The test atmosphere is usually generated by slowly passing a swab wet with the chemical around the sealing surface of the respirator (Figure 3.6). The test atmosphere challenges each general sealing area of the respirator during the inhalation cycle of respiration—the time when leakage is most likely to occur. To prepare for this test, the respirator must be supplied or equipped with a purifying element capable of removing organic vapors.

The major limitation of this method is that it depends on human senses (smell), and odor thresholds vary widely among individuals.

Note: This protocol is not appropriate to use for fit testing particulate respirators. If used to fit test particulate respirators, the respirator must be equipped with an organic vapor filter.

Odor threshold screening, performed without wearing a respirator, is intended to determine if the individual tested can detect the odor of isoamyl acetate at low levels.

Saccharin Test

This method, developed by 3M, may be used on any air-supplied or air-purifying respirator equipped with any air-purifying element approved to remove particulate hazard. The entire screening and testing procedure should be explained to the test subject before conducting the screening test. The saccharin taste threshold

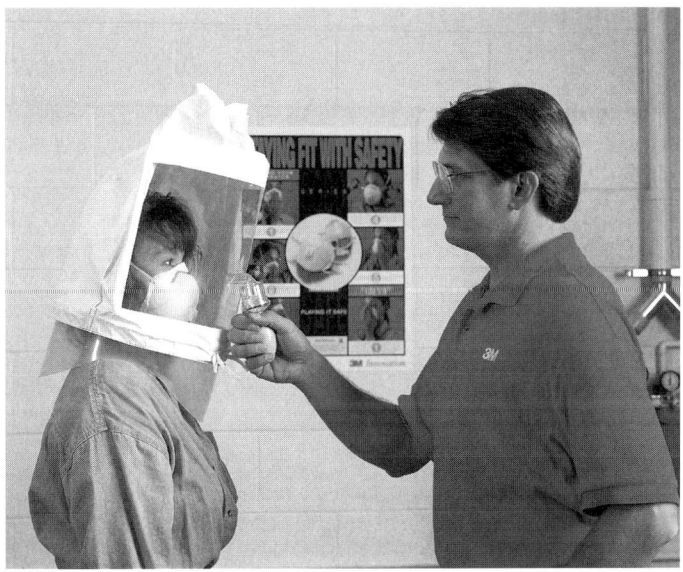

Figure 3.6   Isoamyl acetate qualitative face-fit test (courtesy of 3M Canada).

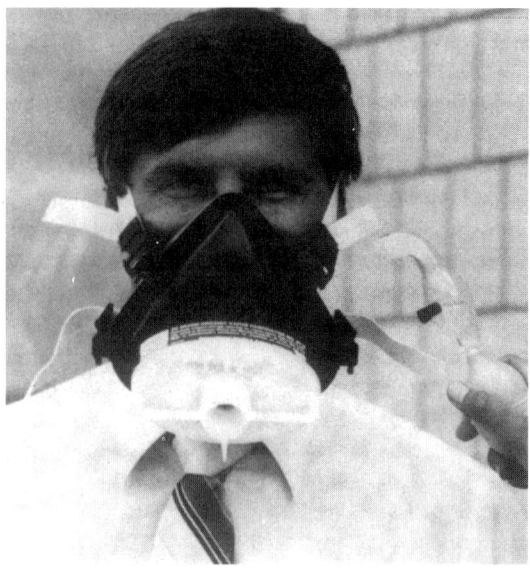

Figure 3.7   Saccharin qualitative face-fit test.

screening, performed without wearing a respirator, is intended to determine whether the individual being tested can detect the taste of saccharin.

The test is generated by nebulizing a saccharin/water solution (Figure 3.7). The water evaporates, leaving a small particle, approximately 2.0 microns in diameter. Again, the sealing edge of the respirator is challenged as before, and the wearer breathes through his mouth throughout the test. If a significant leak is present, the wearer will detect a sweet taste.

If the test subject eats or drinks something sweet before the screening test, he may be unable to taste the weak saccharin solution. As with the isoamyl-acetate method, the major limitation of this method is that it depends on an individual's ability to taste saccharin.

Semi-Quantitative Fit Tests

Semi-quantitative fit tests were designed to offer an alternative to the difficult and costly quantitative fit test methods as they applied to the OSHA Lead Standard. Two methods have been proposed: one designed by DuPont uses isoamyl acetate, and the other, by 3M, uses saccharin.

In principle, both methods are the same. Each one involves first prescreening to determine whether the subject is sensitive to a low, known concentration of the test chemical. If he is sensitive to the concentration, then a qualitative fit test is carried out. If not, then he will not be tested using this method. Therefore, false fits are eliminated.

OSHA published a revision to the fit testing requirements of the Lead Standard Regulations (Standards—29 CFR, **www.osha slc.gov/OshStd_data/1910_1025.html**). In addition to quantitative fit testing, OSHA now permits three methods of qualitative fit testing for assigning respirators to lead-exposed employees. The isoamyl-acetate method, the saccharin-solution-aerosol method, or the irritant-smoke method can be used to test half-mask respirators only. To comply with the Lead Standard, protocol must be followed exactly.

Quantitative Fit Tests

Theoretically, these tests should provide an accurate, objective measurement of fit for all types of respirators. The quantitative fit test is intended to yield a numerical measurement of actual leakage.

Quantitative tests were developed in the early 1960s as a laboratory technique to evaluate respirator fit. Until recently, the equipment and technique were so complex that they remained laboratory tools only. Portable equipment is now available, which has opened the technique to the industrial user.

Two principal methods are currently being used in North America: sodium chloride and dioctyl phthalate (DOP). Other tests use methylene blue, paraffin oil, silica, ethylene gas, or Freon. DOP, the most common test, is used in the form of a polydispersed aerosol as a test agent and is detected with a light-scattering photometer. Sodium chloride is also used in the form of a polydispersed aerosol and is detected with a flame photometer (see Figure 3.8).

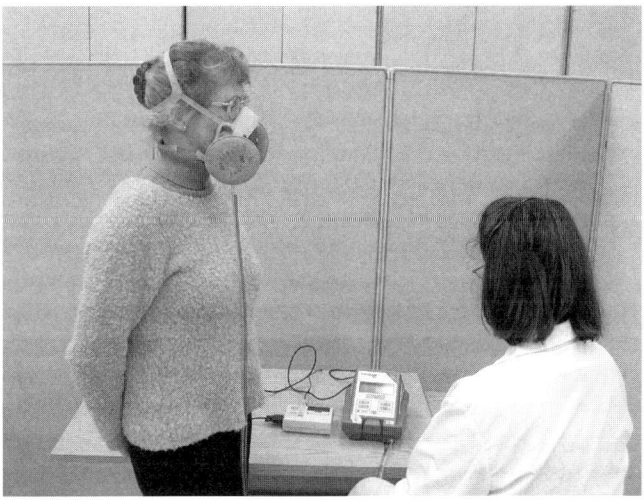

Figure 3.8 Quantitative face-fit test (courtesy of 3M Canada).

OSHA[4] requires the following:

- Quantitative fit testing using a nonhazardous test aerosol (such as corn oil, polyethylene glycol 400 [PEG 400], di-2-ethyl hexyl sebacate [DEHS], or sodium chloride) generated in a test chamber, and employing instrumentation to quantify the fit of the respirator.
- Quantitative fit testing using ambient aerosol as the test agent and appropriate instrumentation (condensation nuclei counter) to quantify the respirator fit.
- Quantitative fit testing using controlled negative pressure and appropriate instrumentation to measure the volumetric leak rate of a facepiece to quantify the respirator fit.

The combination of substitute air-purifying elements, test agent, and test agent concentration shall be such that the test subject is not exposed in excess of an established exposure limit for the test agent at any time during the testing process, based on the length of the exposure and the exposure limit duration.

In order to carry out the quantitative fit tests, the atmosphere is continuously measured inside the respirator through a probe in the respirator facepiece and simultaneously measured outside the respirator (Figure 3.9). The wearer is required to stand in a chamber where the test material atmosphere is generated. The leakage is expressed as a percentage of the test atmosphere outside the respirator. The advantages of quantitative fit testing are as follows:

- Is an excellent training tool to teach the wearer to don, position, and adjust the facepiece for reliable, repeatable, and adequate protection through the instantaneous response of the detector to penetrations of ambient air into the mask.

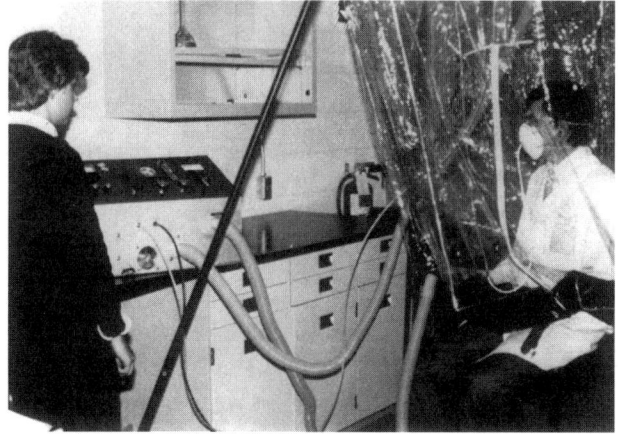

Figure 3.9   Quantitative face-fit test: Portable booth.

- Provides instantaneous feedback of changes in facefit-sealing efficiency caused by movements and adjustments to the facepiece.
- Gives the wearer increased confidence in the assigned respirator.
- Documents results of the respirator-selection process.
- Assesses variations in fit on the individual and over periods of time.
- Reduces the subjectivity in the selection process by generating data for each respirator-wearer and allowing the user to compare results.

The disadvantages of quantitative fit tests are as follows:

- They are very costly. Commercially available portable units range in price from $11,000 to $18,000.
- They are very time-consuming and skill intensive. A skilled operator is required to effectively operate and maintain the equipment. It may take many months of operator training and experience before useful data can be obtained.
- Because the concentration of the test agent inside the respirator varies sinusoidally with the frequency of the breathing rate, and because leakage is usually quantified by peak height, large variations in response time can lead to large variations in measured leakage values.
- Because of the complex nature of quantitative fit test techniques, some people may downplay other aspects of the complete respirator program. Some users may believe that as long as they have data to show that the respirator fits adequately, they may neglect care and maintenance or subject the respirator to rigorous wear.
- The respirator that is tested is not the one worn by workers in the field. It is assumed that different respirators of the same model are similar. This may not always be true because of quality control deficiencies during manufacturing.
- As the name implies, the test determines fit only. It may be dangerous to use quantitative fit tests to derive protection factors.

Once the proper respirator has been selected to match the job hazards and conditions, three essential elements contribute to how well the respirator will protect the user:

- Condition of the equipment
- How well the respirator fits the wearer
- How long the respirator is worn in the contaminated area

The condition of the equipment depends on the quality of the cleaning and maintenance program. If the respirator is not properly maintained, contaminated air can leak into the mask. The fit of the respirator on the wearer's face, as discussed previously, can be expressed as *fit factor*, and the face seal's protection level can be assessed either by qualitative or quantitative methods.

The third consideration involves the worker's willingness to wear the respirator and the time he wears it in the contaminated area. This is considered the *wear factor*, which is a measure of the percent of time the mask is actually worn during the work shift.[26] Even though a respirator has a high fit factor and does an

Table 3.5   Wear factor

FIT FACTOR	80	90	95	100
10	2.80	1.90	1.45	1.00
25	2.32	1.36	0.88	0.40
50	2.16	1.18	0.69	0.20
100	2.08	1.09	0.595	0.10
1000	2.008	1.009	0.509	0.01
INF	2.00	1.00	0.50	0.00

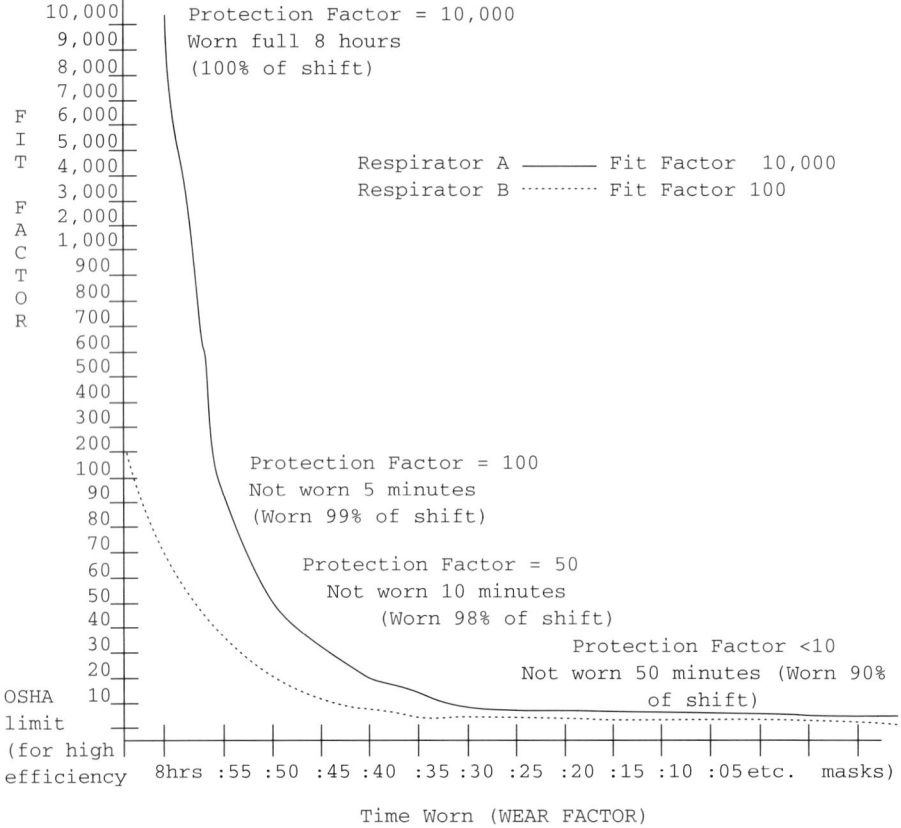

Figure 3.10   Wear factor versus fit factor.

effective job of filtering out the contaminant while being worn, overall protection is drastically reduced by neglecting to wear the respirator in the contaminated area even for brief periods. Figure 3.9 shows how dramatically this protection drops off as the "nonworn" time increases.[27]

The relationship of the fit factor and the wear factor to the protection factor can best be illustrated by examples.

In an eight-hour exposure at 10x PEL, a respirator with fit factor = 1.000 that is worn 95% of the shift (not worn for 24 minutes) will have a protection factor equaling 20. If the respirator is worn 98% of the shift (not worn for 10 minutes), the protection factor is 50. Conversely, a respirator with a fit factor of 100 that is worn 99% of the shift (not worn 5 minutes) will also have a protection factor of 50. Table 3.5 summarizes the wear factor results for different fit factors and percent of time worn. These examples show that keeping a respirator on significantly affects how much protection is provided (Figure 3.10).

The reasons why workers remove their respirators are numerous and include the following: (1) the respirators are heavy and/or uncomfortable; (2) they are difficult to breathe or speak through; or (3) they are too hot. Any selection of a proper respirator must consider these objections in order to ensure a high percentage of respirator-wear time.

CONCLUSION

In summary, many factors must be carefully considered in respirator selection. The nature and extent of the hazard should be determined. The operation and work activities must be studied. The respirator's limitations and characteristics must be weighed with care, taking into account whether IDLH conditions exist and whether the contaminant has poor warning properties. Finally, the respirator must provide the required and proper face seal and degree of protection. All these factors and possibilities must be carefully examined by the health and safety professional when selecting any respiratory protection device. Even when all of these factors are considered, the worker must wear the respirator because when the respirator is removed for a short period, protection is reduced significantly.

REFERENCES

1. OSHA, Code of Federal Regulations, Title 30 Part II, revised July 1, 1986.
2. Bollinger, N.J., and R.H. Schutz. *Guide to Industrial Respiratory Protection.* DHHS (NIOSH) Publication No. 87-116. Cincinnati, OH: NIOSH, 1987. Available at: www.cdc.gov/niosh/87-116.html.
3. Pritchard, J.A. *A Guide to Industrial Respiratory Protection.* DHEW (NIOSH) Publication No. 76-189. Cincinnati, OH: NIOSH, 1976.

4. OSHA, Code of Federal Regulations, Title 42 Part 84, 1996.
5. *NIOSH Certified Equipment List*, as of March 31, 2001, updated periodically. DHHS (NIOSH) Publication No. 2001-139. Cincinnati, OH: NIOSH, 2001. Available at: www.cdc.gov/niosh/87-116.html.
6. Myers, W.R. et al. *NIOSH Respirator Decision Logic*. DHHS (NIOSH) Publication No. 87-108. Cincinnati: OH, NIOSH, 1987.
7. BS EN 372:1992. *Specification for SX gas filters and combined filters against specific named compounds used in respiratory protective equipment*. London, England: British Standards House, 1992.
8. BS EN 371:1992. *Specification for AX gas filters and combined filters against low boiling organic compounds used in respiratory protection equipment*. London, England: British Standards House, 1992.
9. BS EN 143:2000. *Respiratory protective devices. Particle filters. Requirements, testing, marking*. London, England: British Standards House, 2000.
10. BS EN 136:1998. *Respiratory protective devices. Full face masks. Requirements, testing, marking*. London, England: British Standards House, 1998.
11. BS EN 145:1998. *Respiratory protective devices. Self-contained closed-circuit breathing apparatus compressed oxygen or compressed oxygen-nitrogen type. Requirements, testing, marking*. London, England: British Standards House BS, 1998.
12. BS EN 137:1993. *Specification for respiratory protective devices: self-contained open-circuit compressed air breathing apparatus*. London, England: British Standards House, 1993.
13. BS EN 270:1995. *Respiratory protective devices. Compressed air line breathing apparatus incorporating a hood. Requirements, testing, marking*. London, England: British Standards House, 1995.
14. BS EN 400:1993. *Respiratory protective devices for self-rescue. Self-contained closed-circuit breathing apparatus. Compressed oxygen escape apparatus. Requirements, testing, marking*. London, England: British Standards House, 1993.
15. BS EN 12941:1999. *Respiratory protective devices. Powered filtering devices incorporating a helmet or a hood. Requirements, testing, marking*. London, England: British Standards House, 1999.
16. BS EN 149:2001. *Respiratory protective devices. Filtering half masks to protect against particles. Requirements, testing, marking*. London, England: British Standards House, 2001.
17. BS 4275:1997. *Guide to implementing an effective respiratory protective device programme*. London, England: British Standards House, 1997.
18. BS 4400:1969. *Method for sodium chloride particulate test for respirator filters*. London, England: British Standards House, 1997.
19. HSE Guidance HSG53. *Fit testing of respiratory protective equipment used for controlling exposure to asbestos fibres*. London, England: Health and Safety Executive, 1998. Available at: www.hse.gov.uk/lau/lacs/5-18.htm.
20. OSHA. Respiratory Protection Code of Federal Regulations, Title 29 Part 1910.134, 1998.
21. OSHA. Major Requirements of OSHA's Respiratory Protection Standard 29 CFR 1910.134, OSHA Office of Training and Education, March 1998.
22. Canadian Standards Association. Standard CAN/CSA-Z94.4-93. Selection, Use and Care of Respirators, 1997.
23. Pritchard, J.A. *A Guide to Industrial Respiratory Protection*. DHEW (NIOSH) Publication No. 76-189, Appendix F. Cincinnati, OH: NIOSH, 1976.

24. Myers, W.R. "Respiratory Protective Equipment." In R.L. Harris, editor, *Patty's Industrial Hygiene*, 5th ed. New York: John Wiley, 2000; p. 1489.
25. ANSI. "Practice for Respiratory Protection." ANSI Z88.2-1992. New York: American National Standards Institute, 1992.
26. OSHA. "Cadmium" Code of Federal Regulations, Title 29 Part 1910.1027, 1998.
27. 3M Bulletin. *Job Health Highlight*, Vol. 1, No. 3, August 1983.

4

Administration and Training

Chapters 2 and 3 discussed different types of respirators available for use and the selection criteria used to choose the proper respirator. In order to ensure that workers' health is being protected, a respiratory protection program must be established. This program is discussed in more detail in Chapter 7. This chapter discusses two important components of the total program: administration and training.

A respiratory program is needed in workplaces where respirators are used. Such a program establishes responsibilities of the employer, supervisors, and employees and step-by-step procedure for selection, use, and maintenance of respirators. In order for the respiratory program to be effective, administration of the total system and training of all people covered by the program must play a major role. Someone must be put in charge of the implementation and continuous administration of the respirator program. Respirators are misused and will continue to be misused unless an effective training component is incorporated into the program.

The Occupational Safety and Health Administration (OSHA) respirator standard[1] specifies that a respirator program must be administered by a person qualified by appropriate training or experience.

ADMINISTRATION

In order that a respirator program be properly established—started on the right foot—and be effective on an ongoing basis, written standard operating procedures must be established. These procedures should contain all the necessary information needed to maintain an effective respirator program and should be tailored to the users' requirements. The format of written procedures will vary, and should vary, because the requirements in different facilities vary. The written standard operating procedures should address the following topics:

- Guidance for selection of the appropriate respirator, including worksite-specific requirements
- Detailed procedures for:
  - Cleaning, disinfecting, and storing
  - Inspecting, repairing, and replacing

- Fit testing by the wearer and the tester (NIOSH recommends that whenever possible quantitative evaluation of the protection factor should be performed to confirm the actual degree of protection.)
- Purchasing and inventory control of proper respirators
- Issuing of correct respirators
- Ensuring adequate air quality, quantity, and flow of breathing air for atmosphere-supplying respirators
- Training of the wearer, supervisor, administrator, and other appropriate personnel
- Instructions for proper use of respirators in emergencies
- Guidelines for medical surveillance of workers
- Record keeping
- Procedures for evaluating the respirator program's effectiveness

In order to ensure an effective respiratory protection program and that the procedures are written and adhered to, responsibility for the entire program must be assigned to one person, a program administrator. OSHA regulations[1] also recognize that it is essential to clearly identify an individual responsible for administering the respirator program. The exact person to whom this responsibility will be assigned will vary from one organization to another. In a large organization, personnel from several departments (such as Safety and Security, Maintenance, Industrial Relations, Industrial Hygiene) may be involved in the program, but all should report to one administrator. In a small shop, the administrator may be the only person overseeing the program.

The program administrator must have sufficient knowledge, gained by training and experience, to develop and implement the respiratory protection program. He may have a background in safety, industrial hygiene, health care, or engineering. OSHA and other safety and health regulatory bodies around the world require the program administrator to be adequately qualified by training and experience in the proper selection, use, and maintenance of respirators.

The major role of the administrator is to supervise and coordinate. He is responsible for written procedures and their revision when changes take place in the type of hazard present, the process, and so forth. The administrator must be knowledgeable about state-of-the-art respiratory protection and must change the program if needs change. Essentially, the administrator should perform the following tasks:

- Establish a respirator program that meets the needs of the employer and employees.
- Ensure that it is carried out satisfactorily.
- Ensure that it remains effective by continued examination and modification where needed.

Once the respirator program has been established, it must be evaluated periodically (at a minimum, annually) to ensure adequate protection of the worker.

Frequency and conditions that trigger an evaluation must be clearly identified in the program document. This evaluation should be carried out by the administrator or someone outside of the program who is knowledgeable about respiratory protection. Written procedures should be reviewed during the evaluation and changed where necessary. Operation of the program should be reviewed to ensure that (1) proper types of respirators are issued and used; (2) the respirators are properly worn and maintained; (3) the respirators are inspected; and (4) the respirators are properly stored.

During the evaluation, the wearer should be consulted and wearer acceptance should be determined. Comfort, resistance to breathing, fatigue, interference with vision, interference with communications, restrictions of movement, interference with job performance, and confidence in respirator effectiveness should be discussed with the wearer.

Biological and medical monitoring, where applicable, may be used to evaluate the effectiveness of the respirator and, where such data is available, should be reviewed during the evaluation process.

The results of the program evaluation should be presented in a written report and, if modifications are necessary, the plan of action to be implemented must also be included in the report. This report should become part of the employer's record keeping. This record will facilitate the employer not only in maintaining the adequacy of the program but also in determining compliance.

TRAINING

Selecting the proper respirator is one of the most important steps in ensuring the protection of workers in a contaminated environment. Giving the respirator to the worker without teaching him how to use it reduces the probability of proper protection. A training program must be an integral part of any respiratory protection program. This applies to any size facility, whether only several people wear respirators occasionally or a thousand workers wear them daily. Workers should not be expected to know what to do with a respirator, even the simplest model, unless they are trained in its use.

If a respirator is not worn in the contaminated area for even a short period, then the total protection offered by the respirator is drastically reduced. The role of a training program is to educate the wearer about why the respirator is needed and how to wear it properly. It is therefore necessary that the training program be based on the worker's educational level and language background and that it be provided before the wearer uses the respirator.

Table 4.1 shows the probability of respirators being worn given the approach of the employer to the program implementation. Employee education is vital to a successful program, and this factor cannot be downplayed.

Table 4.1   Respirator use success

METHOD OF IMPLEMENTATION	RESULTING SUCCESS PREDICTION	MEANING
By decree	Doubtful	Open to question, invites failure
By persuasion	Dubious	Hesitation, mistrust
By incentive	Perhaps	Possibly but not certainly and may be misunderstood
By education	Probably	Supported by evidence strong enough to establish presumption but not proof
By all of the above	Likely	Having a high probability of occurring

Who should be trained? The following people should be put through the respiratory protection training program:

- The wearer
- The supervisor of the wearer
- The person issuing respirators
- The person performing the fit testing
- The person maintaining and repairing the respirators

Not all of these people need to be trained in all aspects of respiratory protection, but certainly the wearer and supervisor should be exposed to the entire program.

The wearer should receive the most thorough training. This training should convey a clear picture to the wearer about why he should wear a respirator. At the end of the training session, the wearer should be convinced that he needs to wear and maintain the respirator to protect his health. The wearer training should cover the following topics at a minimum:

- The nature, extent, and effect of respiratory hazards to which the person is being exposed
- An explanation about the reasons why respirators need to be used in place of other control methods
- An explanation of why a particular type of respirator is suitable for a given job
- A discussion of the effectiveness and limitations of the respirator being recommended
- The operation, limitations, and capabilities of the selected respirator(s) and an honest appraisal of what could happen if the proper respirator is not used
- Instructions in procedures for inspecting, donning, wearing and removing, and checking the fit and seal of respirators
- Opportunity in the training session to handle the respirator(s), try it on, check the fit, and remove it
- Ability to recognize medical signs and symptoms that may limit or prevent the effective use of respirators

- Proper maintenance and storage procedures
- Instructions in how to deal with emergency situations
- Explanation of the OSHA requirements regarding the training program[2,3]

The wearer's supervisor also needs to be trained. His training is basically the same as that of the wearer, with the inclusion of the following additional points:

- The structure and operation of the entire respiratory protection program
- The legal requirements pertinent to the use of respirators
- The monitoring of respirator use
- The issuance of respirators
- Procedures for coping with emergencies

The person(s) assigned to the task of issuing respirators should be given adequate training and written instructions to ensure that the correct respirator is issued for each application, in accordance with written standard operating procedures.

The person or persons responsible for performing fit tests should be trained in the proper procedure for carrying out the testing process (see Figure 4.1). The tester should be knowledgeable about the different brands and types of respirators used in the workplace and their specific applications.

Adequate training should be given to the personnel who are assigned the responsibility of maintaining and repairing the respirators.

Retraining

It is essential to retrain all personnel involved in the respirator program. Each person should be retrained annually in the appropriate procedures that he needs to ensure that proper effective protection is being provided. Retraining may also be required if changes in the workplace and/or the type of respirator being used have occurred.

Figure 4.1   Train the trainer (courtesy of North Safety Products).

The administrator must set up a record-keeping system for the training program. This system should include the following records as a minimum:

- Who has been trained
- When they received the training
- What training sections they received
- What results were obtained from their fit testing (Is there any type of respirator they cannot wear and why?)
- Results of any medical testing (may indicate types of respirator they cannot wear due to health reasons). Also will indicate if respirator they are wearing is providing the necessary protection.
- Where the person works
- What hazards the person is exposed to
- What type(s) of respirator the person wears
- How long the person wears the respirator during the work shift

Voluntary Respirator Use

Where respirator use is not required, workers may wish to wear respirators voluntarily. The OSHA standard[1] requires that the wearer must be provided with the basic information contained in Appendix D of 29 CFR 1910.34. Thus, it is essential that voluntary users of respirators should also be included in the company's training program.

CONCLUSION

In summary, in order to ensure the proper setup, implementation, and continuation of an effective respiratory protection program, an administrator must be appointed and must be responsible for the entire program. Also, the support of upper management must be given to ensure that the program is a living document that can be modified as required to meet the situations where the use of respirators is changed. In order for a respiratory protection program to be successful, employee acceptance of the wearing of respiratory devices is vital. This acceptance deals more with human relations, and education and training become key factors in motivating employees to use respirators correctly.

REFERENCES

1. Respiratory Protection Code of Federal Regulations, Title 29, section 1910.134, pp. 412–437, 1998.
2. Respiratory Protection Federal Register 63:5, pp. 1258–1262, January 8, 1998.
3. Myers, W.R. "Respiratory Protective Equipment." In Robert L. Harris, editor, *Patty's Industrial Hygiene*, 5th ed. New York: John Wiley, 2000; 1489–1550.

5

Maintenance and Care

The number of U.S. workers who rely on respiratory protection ranges from 2.6 to 7 million. An integral part of a good respiratory protection program is the proper maintenance and care of the respirator. The purpose of a respirator is to prevent inhalation of contaminant air. A serious problem may exist if a wearer dons a poorly maintained or malfunctioning respirator. Workers can derive a false sense of security from respirator use. They feel protected because the respirator is being worn, when in fact the contaminant is being inhaled and serious consequences may result. It is critically important to monitor and maintain respirators throughout their usage life. The integrity of a new facepiece may not be in question, but constant monitoring of its condition is essential as it is used. The goal is to maintain a respirator in a condition that provides the same protection as it was first manufactured.[2]

Poor maintenance is particularly dangerous for emergency escape and rescue devices because they are used infrequently. Each respirator, regardless of the type, must be properly maintained at the level of its original effectiveness. According to the Occupational Safety and Health Administration (OSHA), the American National Standards Institute (ANSI), and the British Standard BS 4275,[3–5] an acceptable program of maintenance and care must include the following:

- Cleaning and sanitizing
- Inspection, testing, and repair
- Storage
- Record keeping

It is an absolute must that any defective or nonfunctioning respirator be removed from use until it is repaired. If the respirator cannot be repaired, then it must be discarded.

CLEANING AND SANITIZING

Cleaning and sanitizing procedures must be included in the respirator protection program and must be part of the wearer's training program. Regulations and consensus standards[3-5] specify that respirators must be cleaned and disinfected at the following intervals:

- Routinely used respirators (personal issue) should be cleaned and disinfected daily.
- Emergency-use respirators and occasionally used respirators should be cleaned after each use.
- Respirators issued to more than one employee should be cleaned and disinfected before being worn by different individuals.
- Respirators used in fit testing and training should be cleaned and sanitized after each use.

In facilities where wearers are assigned individual respirators and are responsible for their own cleaning and disinfecting, each wearer must be thoroughly trained in cleaning and sanitizing procedures. In large establishments where many respirators are used routinely, it may be practical to establish a centralized respirator-cleaning area staffed by specially trained personnel. In such facilities, a worker may not be required to maintain the respirator, but a briefing on cleaning and disinfecting procedures will assure workers that they are receiving a clean, disinfected, and properly maintained respirator.

A suggested procedure for cleaning and sanitizing is as follows:

1. Remove and inspect all removable components (e.g., filters, cartridges, diaphragms, valves). Repair or discard any defective part.
2. Wash the components in warm water with a mild detergent or with a cleaner recommended by the manufacturer. A stiff bristle brush can be used to aid in removing dirt. Care must be taken not to damage any of the components. The cleaning water should not exceed 43.3 °C to prevent damaging the rubber and plastic in the respirator facepieces.
3. Rinse the components thoroughly in clean, warm (43.3 °C) running water and drain.
4. Immerse the respirator components for two minutes in one of the following sanitizing solutions:
   - Hypochlorite solution (50 ppm of chlorine) made by adding approximately 1 ml of laundry bleach to 1 liter of water at 50 °C.
   - Aqueous solution of iodine (50 ppm of iodine) made by adding approximately 0.8 ml of tincture of iodine to 1 liter of water at 43.3 °C.
5. Rinse all components thoroughly in clean, warm (43.3 °C) running water and drain.
6. Dry components with a clean, lint-free cloth or air-dry.
7. Reassemble facepieces, replacing filters, cartridges, or canisters where necessary.
8. Test the respirator to ensure that all components work properly.

In the case of disposable respirators that are reusable, the cleaning and sanitizing procedure is somewhat different and easier because no disassembling and reassembling is involved. If the disposable respirator is to be kept for reuse, then the faceseal area must be wiped with water and a mild detergent or an alcohol wipe. It is then dried and stored in a clean, dry area. The disposable respirator can be considered essentially maintenance free and discarded after its intended use.

INSPECTION, TESTING, AND REPAIR

The wearer must routinely inspect his respirator before and after each use. Manufacturers' instructions for inspection, testing, and repair must be followed to ensure that respirators continue to provide adequate protection. This inspection should include checking the following:

- Tightness of the connections
- Condition of component parts (e.g., facepieces, connecting tubes, filters, cylinders, head harness, valves) for:
  - Presence of excessive dirt
  - Cracks, tears, holes, or distortion from improper storage
  - Cracked or badly scratched lenses in full-facepieces
- End-of-service-life indicator (ESLI)
- Shelf-life dates
- Proper functioning of regulators, alarms, and other warning systems
- Self-contained breathing appartus (SCBA) cylinders filled to specific working pressure
- Pliability and deterioration of rubber or other elastomeric parts

Respirators used for emergency, rescue, or nonroutine purposes should be inspected at least monthly. All respirators should be inspected after being cleaned and sanitized. This will involve the inspection points as outlined previously and the necessary testing to ensure that the respirator and its components are in proper working condition and no detergent residue is left by inadequate rinsing. This residue can cause valve leakage or sticking.

All damaged or faulty components must be repaired or replaced. When this is necessary, the repair and replacement should be performed by persons trained in proper respirator assembly. The proper parts should be used, and the repair and replacement should be carried out as per the manufacturer's instructions. Substitution of parts from one brand or type of respirator to another invalidates NIOSH approval (or equivalent approvals) of the respirator.

OSHA requires that "all respirators should be inspected before and after each use".[6] OSHA further states that respirators that are not used routinely (i.e., emergency escape and rescue devices) "shall be inspected after each use and at least monthly." NIOSH recommends that all stored SCBA be inspected weekly.[6]

STORAGE

All respirators must be properly stored in order to protect them from the environment and the possibility of damage. This includes protection from dust, sunlight, extreme temperature (hot or cold), excessive moisture, and damaging chemicals. All types of respirators should be kept in a cabinet separate from the work environment in a clean-air area. There must be easy access to the storage cabinets for nonroutine or emergency respirators in order to ensure that the wearer is able to put on the necessary equipment to protect himself in case of an emergency. These storage cabinets should be clearly marked. Routinely used respirators should not be stored in workbenches, toolboxes, or lockers unless they are protected from contamination, distortion, and damage. Respirators for emergency use should be:

- Kept accessible to the work area.
- Stored in compartments or in covers that are clearly marked "Emergency Respirators."
- Stored in accordance with the manufacturer's instructions.

RECORD KEEPING

Record keeping is essential to a good maintenance program for respirators. A record of all inspections performed on respirators should be maintained and should include the following:

- Date of use
- Date of inspection
- Physical condition of the respirator
- Date of cleaning and sanitizing of respirator
- Details and dates of repairs carried out
- The tests performed and any corrective action taken, including the dates

These records should be maintained by the administrator of the respiratory protection program.

CONCLUSION

In summary, proper maintenance and care of respirators and their components are important tasks in ensuring the protection of workers. A good respiratory protection program must contain provisions for cleaning, disinfecting, inspecting, and maintaining all respirators covered by the program. Failure to do so may jeopardize the health of workers.

REFERENCES

1. Brosseau, L.M., and Kris Traubel. "An Evaluation of Respirator Maintenance Requirements," *AIHA Journal* 58:116–120, 1997.
2. Rajhans, G.S., and D.S. Blackwell. "A Respiratory Protection Program: The Essential Components," *Applied Industrial Hygiene* 4(11):F-22–F-27, 1989.
3. Occupational Health and Safety Administration. *Inspection Procedures for the Respiratory Protection Standard*, CPL 2.120. Washington, DC: U.S. Department of Labor/ OSHA, September 18, 1998.
4. American National Standards Institute (ANSI). *American National Standard for Respiratory Protection*. New York: ANSI, Z88.2 1992; 16–17.
5. British Standards Institute. BS 4275:1997, *Recommendations for the Selection, Use and Maintenance of Respiratory Protective Equipment*. London, England: BSI, 1997.
6. Bollinger, N.J., and R.H. Schutz. *NIOSH Guide to Respiratory Protection*, DHHS (NIOSH) Publication No. 87-116. Cincinnati, OH: NIOSH, 1987.

6

Medical Supervision

The use of a respirator, regardless of the type, imposes some physical discomfort and distress. The degree of distress depends on various factors such as:

- *Breathing resistance*. For example, air-purifying respirators make breathing more difficult. The special exhalation valve on an open-circuit pressure-demand respirator requires the wearer to exhale against significant resistance.
- *Vision limitation*. Vision is restricted to some extent with all devices, which can lead to tunnel vision or perceptual narrowing. This response can, in turn, lead to anxiety in certain individuals above and beyond the stress induced by heavy physical work (e.g., firefighting).
- *Movement restriction*. Movement is restricted because of the cumbersome nature of some equipment. For example, if the wearer is using an air-line respirator, he might have to drag up to 300 feet of hose around.
- *Communication restriction*. All respiratory equipment restricts communication. This can become unpleasant and, at times, add to existing stress.
- *Weight*. All respiratory devices impose an extra burden on the wearer, but some heavier devices, such as a self-contained breathing appartus (SCBA), add even greater strain.
- *Thermal stress*. Covering the face with an impervious mask significantly impairs heat loss from the head. Depending on the work activity, the sweat rate may increase by as much as 8%, but evaporation sweat loss may be reduced by 12%.[1]
- *Psychophysiological effects*. A feeling of claustrophobia may develop in some wearers of respirators. Identifying those individuals who experience such problems is not always easy.

Now let us consider a worker suffering from emphysema who is required to wear air-purifying devices. It is not hard to visualize the severe complications he would develop. Similarly, diseases such as asthma and chronic bronchitis will severely limit respirator usage. Those who suffer from heart disease may find that any of the aforementioned factors accentuate their problems. Epileptic workers may rip off their masks while working in atmospheres that are immediately dangerous to life. It is therefore imperative to determine individual fitness or ability to wear respirators without aggravating a preexisting medical problem. The following guidelines will assist in developing an effective medical surveillance program.

THE EXAMINING PHYSICIAN

The three main purposes of medical supervision are as follows:

1. To comply with legislative requirements.
2. To determine fitness to work in the workplace.
3. To follow changes in the health status of employees over time and assess if a correlation exists between those changes and the working environment.

The examining physician has the responsibility for directing an effective health-surveillance medical program with major emphasis on prevention. In order to render a qualified opinion regarding respirator usage by an employee, the physician initially should obtain the following information:

● Type of respiratory protection to be used and its modes of operation
● The contaminant(s) to which the employee will be exposed and related toxicity data
● Tasks that the employee will perform while wearing the respirator
● Visual and audio requirements associated with the task
● Length of time the user will wear the respiratory equipment
● An estimation of the number of emergency or rescue situations
● Length of time the user may be required to spend in extreme environments

The preceding information, obtained from plant management or a plant industrial hygienist, can be used to develop effective programs and procedures that are appropriate to specific operating locations and their conditions. At a minimum, the program should consist of the following:

● Preemployment or job change medical examination
● Routine periodic medical examination
● Examination on return to work following injury or prolonged illness
● Medical records

PREEMPLOYMENT OR JOB CHANGE EXAMINATION

The preemployment medical examination is a screening mechanism for respirator usage. At the end of this examination, the physician must be able to determine whether the worker is physically or mentally suitable for jobs requiring the use of respirators and whether there are any particular jobs to which the worker should not be assigned or posts for which he is better suited. The examination should include the following parameters:

History

Taking a history involves collecting data on the worker's past occupational exposure, present or past respiratory disorders, and personal habits (smoking and hygiene).

Questionnaire

Appendix C of the OSHA Standard 1910.134[1] gives the detailed medical evaluation questionnaire (reproduced at the end of this chapter). The American National Standards Institute (ANSI) Publication Z 88.6[2] also discusses medical examination guidelines for employees using respiratory equipment.

Our review of all the federal and state regulations indicates that, as a minimum, the following questions should be included on a questionnaire designed to identify any circulatory or respiratory system medical problems that can contribute to an employee's decreased or regulated use of a respirator:

1. Have you ever been short of breath on exertion, such as (a) climbing a flight of stairs, (b) walking up a slight hill, or (c) walking with other people of your own age, at an ordinary pace on level ground?
2. Have you ever had chest pains?
3. Have you ever had asthma or emphysema?
4. Do you frequently have difficulty breathing?
5. Do you smoke?
6. Have you ever had a heart attack or other heart condition?
7. Do you have chronic skin problems of the face?
8. Do you wear glasses or contact lenses?
9. Do you have high blood pressure?
10. Do you know of any reason why you are not able to wear a respirator?

Any employee who has a past history of respiratory problems should be forbidden to wear a respirator that restricts inhalation and exhalation. For example, a person suffering from emphysema may be unable to breathe adequately against the additional resistance of a respirator. Similarly, if the user suffers an asthma attack, he would be likely to remove the respirator because he would be unable to breathe properly.

Details of previous workplace exposure must also be obtained during the initial medical examination. A record should be kept of jobs the worker has had since leaving school, on what conditions and for what duration, as well as any occupational accidents and diseases for which he may have been compensated. This information is useful because many occupational diseases, such as asbestosis, silicosis, or occupational tumors, often have a latent period.

General Physical Examination

In physical examinations, particular attention should be directed toward physical and sensorial characteristics and those systems that may be affected by wearing respirators. The essential physical characteristics that should be noted are excessive weight, scars, hollow temples, very prominent cheekbones, deep skin creases, and lack of teeth or dentures, which may cause respirator facepiece-sealing problems and inability to use fingers or hands. Respirators such as gas masks, supplied-air respirators, and SCBA require connection and disconnection of parts and manipulation of valves and fittings during use. Persons with missing or disabled fingers may have difficulty in using these devices, particularly in an emergency when there may be no one present to assist them.

The significant sensorial characteristics to be noted are poor eyesight and poor hearing. These defects, if undetected, may increase the on-the-job risks involved in using certain types of respirators.

The correct functioning of the various systems of the body should also be checked. For example, the presence of a hernia can be aggravated by wearing/carrying respiratory equipment (SCBA). Temporary congestion caused by seasonal allergies (such as hayfever) may make it unpleasant to work with a respirator on.

Medical and Psychological Tests

Tests should be performed to examine the major systems. Pulmonary function tests should be undertaken in conjunction with x-rays and should include measurement of forced vital capacity (FVC) and forced expiratory volume at one second (FEV_1). These tests will reveal respiratory insufficiency, including asthma, chronic bronchitis, or emphysema. X-rays assist in determining the morphologic state of the heart and its vascular pedicle. Blood and urine tests should be as extensive as necessary to detect anemia, hemophilia, and disorders resulting from occupational poisoning, or increased absorption of toxic substances (e.g., lead and mercury). Thus, medical tests must be general as well as specific to the substance to which the employee will be exposed.

As stated earlier, some individuals may experience phobic sensations or be anxious because of a conscious appreciation of breathing. Unfortunately, there is no diagnostic tool in use by which an occupational physician can screen these individuals. Hence psychological screening may be subjective; however, when respirators are required to be used in emergency or rescue situations, the physician must take mental factors into consideration before recommending that psychologically affected individuals be allowed to work in situations where panic may endanger not only their lives, but their coworkers' as well.

PERIODIC EXAMINATION

Periodic examinations are traditionally designed to detect and treat potentially dangerous disease, identify risk factors, and prolong life through risk-factor modification or early treatment. For respirator users, there are some benefits of periodic examinations, such as finding out what problems workers are having, answering their questions factually, educating workers on prudent work practices and respirator use, informing them of their exposure to toxicants, and reassuring them through candid discussions.

The periodic health evaluation ascertains that workers' capacities, as defined during the preemployment examination, remain compatible with their occupation, as well as ensures that the work they carry out does not cause any additional disorder or lesion. Risk-factor intervention may appreciably improve the quality of life even though the length of life may not be increased. Workers who are successful in modifying harmful health practices usually derive impressive benefits. During these evaluations, much valuable health-experience information can be accumulated for future epidemiologic studies. In addition, sudden temporary incapacitation or gradual deterioration of sensory facilities, alertness, or agility can also be detected.

The frequency of examinations will depend on the age and health of workers and the risk factors associated with their jobs. For example, workers in excellent health with few or no risk factors and little exposure to toxic substances certainly need to be examined less frequently than those under multiple risks. NIOSH Respirator Decision Logic[3] suggests a medical monitoring frequency as shown in Table 6.1.

EXAMINATION ON RETURN TO WORK AFTER INJURY OR PROLONGED ILLNESS

This evaluation should ensure that the worker whose work requires certain protective equipment is capable of resuming his usual work. Any changes in work capacity that may have been caused by the illness or accident should also be looked for because permanent physical handicaps may make resumption of the former work impossible (occupational handicap). It then becomes necessary to reassign the worker.

An attempt should be made to find out if any connection exists between the injury and the worker's job in order to improve preventive measures. It may be

Table 6.1   Suggested frequency of medical monitoring worker-age

CONDITION	< 35	35–45	> 45
Most work condition requiring respirators	Every 5 years	Every 2 years	1–2 years
Strenuous work conditions with SCBA	Every 3 years	Every 18 months	Annually

necessary for the physician to visit the workplace, investigate the environmental conditions, and determine the precise nature of the substances used. Further questioning of the worker may reveal some valuable information regarding the working environment, the tempo of work, the efficiency of engineering controls, and the use of various protective devices available to him.

If the worker resists using prescribed respiratory equipment, the reasons must be accurately determined by questioning him about all the operations he has to perform, their duration, their frequency, and any technical changes that may have been introduced since he began his work. It may also be useful to ascertain relationships with his supervisors, coworkers, and subordinates.

MEDICAL RECORDS

Medical records of a worker must be maintained for ready access to the documents relating to preemployment examinations, periodic checkups, the results of specialist examinations, x-ray or laboratory tests, individual accident report forms, and exposure records. These records should be maintained in a confidential file for use by the physician or nurse because, based on these confidential records, the physician concludes that a worker is fit, fit with limitations, or unfit to perform certain duties, and advises the employer accordingly, without revealing the details of personal records. Medical ethics require that complete observance of professional confidentiality be maintained. In some countries, the examining physician is required by law to make certain disclosures of medical records to the worker or his union, and in some instances to judicial authorities. The necessary information can usually be given without impairing professional secrecy by limiting the disclosure of information to only such facts as are absolutely essential.

ACTIONS TO BE TAKEN ON MEDICAL INFORMATION

If a physician advises that a worker is fit with limitations, or unfit to wear respirators, the employer must act on this information. The precise action taken will depend on the professional advice of the physician. In many cases, there may be some choice regarding the course of action taken, which can be worked out between the physician, the employer, and the worker.

If the results of clinical tests reach a critical level (e.g., .30 µg/100 ml of blood lead), then the worker must be removed from exposure to the contaminant until results reach acceptable levels. High test results may indicate a lack of protection afforded by the use of respirators. A thorough investigation of the conditions of work and work practices of the worker must then be made. The effectiveness of respirators under present conditions of work should also be examined. It is possible that conditions have changed significantly to warrant better respiratory protection, or better engineering controls, in order to reduce contaminant levels. Conversely, the results of clinical tests may not reach a critical level, but the physician may still advise that a worker is fit with limitations or unfit to wear respirators on the basis of other signs of adverse health effects. The employer should adhere to the advice and information that the physician has provided because it aids in evaluating and improving workplace controls. Although it may be necessary, in some cases, to provide more efficient respiratory protection, the best course of action is to reduce environmental contamination so that the worker does not have to be retrained to use new equipment.

CONCLUSION

There are physiological and psychological problems associated with the use of respirators. These problems are further compounded if the user feels that his health may be compromised. For these reasons, a medical evaluation questionnaire followed by a medical evaluation is of paramount importance. Regular medical supervisions will ensure that additional medical evaluation is conducted whenever the user reports a medical sign or symptom that may be related to the use of a respirator.

REGULATIONS (STANDARDS-29 CFR)

OSHA Respirator Medical Evaluation Questionnaire (Mandatory)— 1910.134AppC

OSHA Regulations (Standards-29 CFR)—Table of Contents

- **Standard Number**: 1910.134AppC
- **Standard Title**: OSHA Respirator Medical Evaluation Questionnaire (Mandatory).
- **SubPart Number**: I
- **SubPart Title**: Personal Protective Equipment

Appendix C to Sec. 1910.134: OSHA Respirator Medical Evaluation Questionnaire (Mandatory)

To the employer: Answers to questions in Section 1, and to question 9 in Section 2 of Part A, do not require a medical examination.

To the employee:
Can you read (circle one): Yes/No

Your employer must allow you to answer this questionnaire during normal working hours, or at a time and place that is convenient to you. To maintain your confidentiality, your employer or supervisor must not look at or review your

answers, and your employer must tell you how to deliver or send this questionnaire to the health care professional who will review it.

Part A. Section 1. (Mandatory) The following information must be provided by
every employee who has been selected to use any type of respirator (please print).

1. Today's date: _____
2. Your name: _____
3. Your age (to nearest year): _____
4. Sex (circle one): Male/Female
5. Your height: _____ft. _____in.
6. Your weight: _____lbs.
7. Your job title: _____
8. A phone number where you can be reached by the health care professional
    who reviews this questionnaire (include the Area Code): _____
9. The best time to phone you at this number: _____
10. Has your employer told you how to contact the health care professional who
    will review this questionnaire (circle one): Yes/No
11. Check the type of respirator you will use (you can check more than one
    category):
    a. ___ N, R, or P disposable respirator (filter-mask, noncartridge type only).
    b. ___ Other type (for example, half- or full-facepiece type, powered-air
          purifying, supplied-air, self-contained breathing apparatus).
12. Have you worn a respirator (circle one): Yes/No

    If "yes," what type(s): _____

Part A. Section 2. (Mandatory) Questions 1 through 9 below must be answered
by every employee who has been selected to use any type of respirator (please
circle "yes" or "no").

1. Do you **currently** smoke tobacco, or have you smoked tobacco in the last
    month: Yes/No
2. Have you **ever had** any of the following conditions?
    a. Seizures (fits): Yes/No
    b. Diabetes (sugar disease): Yes/No
    c. Allergic reactions that interfere with your breathing: Yes/No
    d. Claustrophobia (fear of closed-in places): Yes/No
    e. Trouble smelling odors: Yes/No
3. Have you **ever had** any of the following pulmonary or lung problems?
    a. Asbestosis: Yes/No
    b. Asthma: Yes/No
    c. Chronic bronchitis: Yes/No
    d. Emphysema: Yes/No
    e. Pneumonia: Yes/No
    f. Tuberculosis: Yes/No
    g. Silicosis: Yes/No

    h. Pneumothorax (collapsed lung): Yes/No
    i.   Lung cancer: Yes/No
    j.   Broken ribs: Yes/No
    k. Any chest injuries or surgeries: Yes/No
    l.   Any other lung problem that you've been told about: Yes/No

4. Do you **currently** have any of the following symptoms of pulmonary or lung illness?
    a. Shortness of breath: Yes/No
    b. Shortness of breath when walking fast on level ground or walking up a slight hill or incline: Yes/No
    c. Shortness of breath when walking with other people at an ordinary pace on level ground: Yes/No
    d. Have to stop for breath when walking at your own pace on level ground: Yes/No
    e. Shortness of breath when washing or dressing yourself: Yes/No
    f.   Shortness of breath that interferes with your job: Yes/No
    g. Coughing that produces phlegm (thick sputum): Yes/No
    h. Coughing that wakes you early in the morning: Yes/No
    i.   Coughing that occurs mostly when you are lying down: Yes/No
    j.   Coughing up blood in the last month: Yes/No
    k. Wheezing: Yes/No
    l.   Wheezing that interferes with your job: Yes/No
    m. Chest pain when you breathe deeply: Yes/No
    n. Any other symptoms that you think may be related to lung problems: Yes/No

5. Have you **ever had** any of the following cardiovascular or heart problems?
    a. Heart attack: Yes/No
    b. Stroke: Yes/No
    c. Angina: Yes/No
    d. Heart failure: Yes/No
    e. Swelling in your legs or feet (not caused by walking): Yes/No
    f.   Heart arrhythmia (heart beating irregularly): Yes/No
    g. High blood pressure: Yes/No
    h. Any other heart problem that you've been told about: Yes/No

6. Have you **ever had** any of the following cardiovascular or heart symptoms?
    a. Frequent pain or tightness in your chest: Yes/No
    b. Pain or tightness in your chest during physical activity: Yes/No
    c. Pain or tightness in your chest that interferes with your job: Yes/No
    d. In the past two years, have you noticed your heart skipping or missing a beat: Yes/No
    e. Heartburn or indigestion that is not related to eating: Yes/ No
    f.   Any other symptoms that you think may be related to heart or circulation problems: Yes/No

7. Do you **currently** take medication for any of the following problems?
    a. Breathing or lung problems: Yes/No

    b.   Heart trouble: Yes/No
    c.   Blood pressure: Yes/No
    d.   Seizures (fits): Yes/No

8.   If you've used a respirator, have you **ever had** any of the following problems? (If you've never used a respirator, check the following space and go to question 9:)
    a.   Eye irritation: Yes/No
    b.   Skin allergies or rashes: Yes/No
    c.   Anxiety: Yes/No
    d.   General weakness or fatigue: Yes/No
    e.   Any other problem that interferes with your use of a respirator: Yes/No

9.   Would you like to talk to the health care professional who will review this questionnaire about your answers to this questionnaire: Yes/No

Questions 10 to 15 below must be answered by every employee who has been selected to use either a full-facepiece respirator or a self-contained breathing apparatus (SCBA). For employees who have been selected to use other types of respirators, answering these questions is voluntary.

10.   Have you **ever lost** vision in either eye (temporarily or permanently): Yes/No

11.   Do you **currently** have any of the following vision problems?
    a.   Wear contact lenses: Yes/No
    b.   Wear glasses: Yes/No
    c.   Color blind: Yes/No
    d.   Any other eye or vision problem: Yes/No

12.   Have you **ever had** an injury to your ears, including a broken ear drum: Yes/No

13.   Do you **currently** have any of the following hearing problems?
    a.   Difficulty hearing: Yes/No
    b.   Wear a hearing aid: Yes/No
    c.   Any other hearing or ear problem: Yes/No

14.   Have you **ever had** a back injury: Yes/No

15.   Do you **currently** have any of the following musculoskeletal problems?
    a.   Weakness in any of your arms, hands, legs, or feet: Yes/No
    b.   Back pain: Yes/No
    c.   Difficulty fully moving your arms and legs: Yes/No
    d.   Pain or stiffness when you lean forward or backward at the waist: Yes/No
    e.   Difficulty fully moving your head up or down: Yes/No
    f.   Difficulty fully moving your head side to side: Yes/No
    g.   Difficulty bending at your knees: Yes/No
    h.   Difficulty squatting to the ground: Yes/No
    i.   Climbing a flight of stairs or a ladder carrying more than 25 lbs: Yes/No
    j.   Any other muscle or skeletal problem that interferes with using a respirator: Yes/No

Part B Any of the following questions, and other questions not listed, may be added to the questionnaire at the discretion of the health care professional who will review the questionnaire.

1. In your present job, are you working at high altitudes (over 5000 feet) or in a place that has lower than normal amounts of oxygen: Yes/No

    If "yes," do you have feelings of dizziness, shortness of breath, pounding in your chest, or other symptoms when you're working under these conditions: Yes/No

2. At work or at home, have you ever been exposed to hazardous solvents, hazardous airborne chemicals (e.g., gases, fumes, or dust), or have you come into skin contact with hazardous chemicals: Yes/No

    If "yes," name the chemicals if you know them: _____
    _____
    _____

3. Have you ever worked with any of the materials, or under any of the conditions, listed below:
    a. Asbestos: Yes/No
    b. Silica (e.g., in sandblasting): Yes/No
    c. Tungsten/cobalt (e.g., grinding or welding this material): Yes/No
    d. Beryllium: Yes/No
    e. Aluminum: Yes/No
    f. Coal (for example, mining): Yes/No
    g. Iron: Yes/No
    h. Tin: Yes/No
    i. Dusty environments: Yes/No
    j. Any other hazardous exposures: Yes/No

    If "yes," describe these exposures:_____
    _____
    _____

4. List any second jobs or side businesses you have: _____
    _____
    _____

5. List your previous occupations: _____
    _____
    _____

6. List your current and previous hobbies: _____
    _____
    _____

7. Have you been in the military services? Yes/No

    If "yes," were you exposed to biological or chemical agents (either in training or combat): Yes/No

8. Have you ever worked on a HAZMAT team? Yes/No
9. Other than medications for breathing and lung problems, heart trouble, blood pressure, and seizures mentioned earlier in this questionnaire, are you taking any other medications for any reason (including over-the-counter medications): Yes/No

    If "yes," name the medications if you know them:_____

10. Will you be using any of the following items with your respirator(s)?
    a. HEPA Filters: Yes/No
    b. Canisters (for example, gas masks): Yes/No
    c. Cartridges: Yes/No
11. How often are you expected to use the respirator(s) (circle "yes" or "no" for all answers that apply to you)?
    a. Escape only (no rescue): Yes/No
    b. Emergency rescue only: Yes/No
    c. Less than 5 hours **per week**: Yes/No
    d. Less than 2 hours **per day**: Yes/No
    e. 2 to 4 hours per day: Yes/No
    f. Over 4 hours per day: Yes/No
12. During the period you are using the respirator(s), is your work effort:
    a. **Light** (less than 200 kcal per hour): Yes/No

        If "yes," how long does this period last during the average shift:_____
        hrs. _____mins.

Examples of a light work effort are **sitting** while writing, typing, drafting, or performing light assembly work; or **standing** while operating a drill press (1–3 lbs.) or controlling machines.

    a. **Moderate** (200 to 350 kcal per hour): Yes/No

        If "yes," how long does this period last during the average shift:_____
        hrs. _____mins.

Examples of moderate work effort are **sitting** while nailing or filing; **driving** a truck or bus in urban traffic; **standing** while drilling, nailing, performing assembly work, or transferring a moderate load (about 35 lbs.) at trunk level; **walking** on a level surface about 2 mph or down a 5-degree grade about 3 mph; or **pushing** a wheelbarrow with a heavy load (about 100 lbs.) on a level surface.

a.   **Heavy** (above 350 kcal per hour): Yes/No

If "yes," how long does this period last during the average shift:_____
hrs. _____mins.

Examples of heavy work are **lifting** a heavy load (about 50 lbs.) from the floor to your waist or shoulder; working on a loading dock; **shoveling; standing** while bricklaying or chipping castings; **walking** up an 8-degree grade about 2 mph; climbing stairs with a heavy load (about 50 lbs.).

13.   Will you be wearing protective clothing and/or equipment (other than the respirator) when you're using your respirator: Yes/No

If "yes," describe this protective clothing and/or equipment: _____
_____

14.   Will you be working under hot conditions (temperature exceeding 77°F): Yes/No
15.   Will you be working under humid conditions: Yes/No
16.   Describe the work you'll be doing while you're using your respirator(s): _____
17.   Describe any special or hazardous conditions you might encounter when you're using your respirator(s) (for example, confined spaces, life-threatening gases):_____
18.   Provide the following information, if you know it, for each toxic substance that you'll be exposed to when you're using your respirator(s):

Name of the first toxic substance:_____
Estimated maximum exposure level per shift: _____
Duration of exposure per shift:_____
Name of the second toxic substance: _____
Estimated maximum exposure level per shift: _____
Duration of exposure per shift: _____
Name of the third toxic substance: _____
Estimated maximum exposure level per shift: _____
Duration of exposure per shift: _____
The name of any other toxic substances that you'll be exposed to while using your respirator: _____
_____
_____

19.   Describe any special responsibilities you'll have while using your respirator(s) that may affect the safety and well-being of others (for example, rescue, security):
_____

REFERENCES

1. Respiratory Protection Code of Federal Regulations, Title 29 Part 1910.134, Appendix C, 1998.
2. American National Standards Institute. American National Standard for Respiratory Protection (ANSI Z 88.6), New York: ANSI, 1984.
3. Myers, W.R. et al. *Respirator Decision Logic*. DHHS (NIOSH) Publication No. 87-108. Cincinnati, OH: National Institute for Occupational Health, May 1987.

7

Respiratory Protection Program

This chapter highlights the necessary components of proper and effective respiratory protection programs. With all of the current legislation promulgating the use of engineering controls, work practices, and hygiene practices and facilities to control the exposure of a worker to a particular contaminant, you may wonder why respiratory protection is needed. Regardless of any legislation, the responsibility of every employer should be to safeguard the health, safety, and general well-being of employees. In many cases, this will involve the use of respirators. Respirator use is only one alternative. As any alternative solution to a problem, its use should be determined by considering its risks, practicality, and cost-effectiveness. As a general rule of thumb, respirators should be used in the following conditions:

- Operations where environmental controls are being implemented or evaluated.
- Operations where environmental controls are unavailable or cannot by themselves maintain an acceptable exposure level.
- Operations involving constructions, maintenance, repair, and decontamination where environmental controls may be unavailable or impracticable.
- Emergencies where there is a health risk.

But using respirators opens a Pandora's box of questions regarding the protection we are getting, such as: What type of respirator should we use? Why are we using it? Is it the proper one? Does it fit? The questions go on. A respirator cannot be simply given to a worker with the expectation that it will provide total protection. It must be presented as part of a respiratory protection program. Moreover, a written respiratory protection program should contain worksite-specific procedures and elements for required respirator use. A suitably trained program administrator must administer the program. Previous chapters give more detailed explanations of individual components of such a program.

According to the current OSHA Standard, Section 1910.134, a formal respiratory protection equipment program is necessary when respirators are used in various manufacturing facilities. OSHA Standard 29CFR 1910.134(b) requirements for a minimally acceptable respiratory protection program are as follows:

- Procedures for selecting respirators for use in the workplace
- Medical evaluations of employees required to use respirators
- Fit-testing procedures for tight-fitting respirators
- Procedures for proper use of respirators in routine and reasonably foreseeable emergency situations
- Procedures and schedules for cleaning, disinfecting, storing, inspecting, repairing, discarding, and otherwise maintaining respirators
- Procedures to ensure adequate air quality, quantity, and flow of breathing air for atmosphere-supplying respirators
- Training of employees in the respiratory hazards to which they are potentially exposed during routine and emergency situations
- Training of employees in the proper use of respirators, including putting on and removing them, any limitations on their use, and their maintenance
- Procedures for regularly evaluating the effectiveness of the program
- Procedures for providing respirators at the request of employees or permitting employees to use their own respirators, where respirator use is not required but such respirator use will not in itself create a hazard
- Designated program administrator who is qualified by appropriate training or experience that is commensurate with the complexity of the program to administer or oversee the respiratory protection program and conduct the required evaluations of program effectiveness
- Provision of respirators, training, and medical evaluations at no cost to employees

The first step in setting up a program is to assign a program administrator. The administrator, chosen on the basis of training and experience, is solely responsible for all facets of the program. He must wear many hats in this position of coordinator, educator, supervisor, and consultant. The administrator must be fully aware of the hazards, posted areas, employee limitations, and employee acceptance of respirators. He must also be familiar with all aspects of the employer's and employees' responsibilities as described as follows:

Employer's Responsibilities

- Determination of workplace exposure to hazardous air contaminants
- Respirator fit testing before use
- Random inspections to ensure that:
  - Proper respirators are being used.
  - Respirators are being worn properly.
- Consultation with respirator users about:
  - Breathing difficulties
  - Fatigue

- Interference with vision
- Restriction of movement
- Interference with job performance
- Confidence in the respirator

Employee's Responsibilities

- Check the respirator fit after each donning as instructed.
- Use the respirator as instructed.
- Prevent damage to the respirator.
- Go immediately to an area with respirable air if the respirator fails to provide proper protection.
- Report any damages or malfunction to a person responsible for the respirator program.

The entire respiratory protection program should be outlined in written standard operating procedures. These procedures should be readily available to all people affected, especially the wearer, and cover the following topics:

- Identification of the hazard
- Hazard assessment
- Hazard control
- Respirator selection on the basis of hazards
- Respirator fit
- Training program
- Maintenance procedures
- Medical surveillance
- Program evaluation

The remainder of this chapter expands on each of these nine components.

IDENTIFICATION OF HAZARD

Two basic conditions need to be identified. The first is to identify the areas of the plant where there are problems. This involves in-depth consideration of the process, equipment, and the plant. The second is to identify the type of hazard. For example, when we examine a particulate problem, dusts may be the only form of particulates you may be aware of; however, others exist as well, such as mists, fumes, spray, smoke, fog, smog, and radon daughters. All particulate hazards require varying levels of protection. In general, you can classify your problem into four main categories:

1. Oxygen deficiency
2. Particulates

3.   Gas/vapors
4.   A combination of all of these

This general classification can narrow down the respirator selection procedure; however, as is the case with particulates, an in-depth look at the process and the chemistry involved within these categories is necessary to adequately define the level of protection needed.

HAZARD ASSESSMENT

A large part of the assessment stage is knowledge of the potential toxicity of the contaminant. Considerations include occupational exposure limits, e.g., TLVs, routes of entry, and biological effects. The toxicity and behavior of individual chemicals will play a large part in the choice of a respirator. In order to determine the exposure level, air samples of the workplace should be taken. Samples should represent work periods.

HAZARD CONTROL

Problem areas identified by the assessment require effective engineering controls, such as isolation, substitution, removing the source, and ventilation, in order to reduce worker exposure to respiratory hazards. The purpose of the program is to protect the worker's respiratory system. The most effective way of doing this is by reducing exposure through technological improvements.

Where it is not practical for control procedures to be instituted, or during the time that controls are being installed, respiratory protection appropriate to the circumstances should be worn. Respirators should never be considered a substitute for engineering controls.

RESPIRATOR SELECTION

Selection of the proper respirator is the most complex area of the program. There are many different potential contaminants, as well as many different forms of contaminants, so it follows that there are many different types of respirators as well.

A starting point in respirator selection is to look for an approval or certification label. Remember, however, that the National Institute for Occupational Safety and Health (NIOSH) does not have approval schedules for all contaminants.

There are many different types of respirators to select from. These include self-contained breathing apparatus (SCBA); supplied-air respirators; gas masks; dust, fume, and mist (DFM) respirators; and chemical-cartridge or combination chemical-cartridge respirators. Selection of the proper respirator for a given situation requires considering (1) the nature of the hazard; (2) characteristics of the hazardous operation or process; (3) location of the hazardous area with respect to a safe area having respirable air; (4) period of time in use; (5) nature of the work; (6) and the physical characteristics, functional capabilities, and limitations of various types of respirators. When selecting a respiratory protection device, the administrator must carefully examine all factors and possibilities.

RESPIRATOR FIT

The ability of a respirator to do the job depends particularly on the effectiveness of the respirator itself in a given environment and the proper use and fit of that respirator. It cannot be overemphasized that a proper fit must be obtained when respirators are worn. It is essential that the user knows the capability of the respirator he is required to wear and the degree of protection he can reasonably expect from it. As discussed in Chapter 3, several factors affect respirator fit, including design of the respirator; facial contours and facial hair; training and knowledge required for respirator selection; worker acceptance; and maintenance of respirators.

Respirators should be selected according to the characteristics of the workplace hazards, the capabilities and limitations of the respirator and, most important, the ability of each respirator wearer to obtain a satisfactory fit.

Respirator fit, simply stated, is the ability of the device to interface with the wearer to prevent the workplace atmosphere from entering the worker's respiratory system. When the contaminated atmosphere penetrates the interface and enters the worker's breathing zone, exposure occurs. To effectively use respirators as a means of controlling airborne hazards, assessment and control must be made of the exposure resulting from an imperfect fit.

Several types of fit tests are available. The two main types are the qualitative and quantitative fit tests. The qualitative or quantitative fit test should be used to determine the ability of each individual respirator wearer to obtain a satisfactory fit with a negative-pressure respirator. The results of these fit tests can be used to determine a fit factor.

The fit factor, as discussed in Chapter 3, is a measure of the degree of protection provided by the respirator to the wearer. It is the ratio of the concentration of a substance in ambient air to its concentration inside the respirator when worn. The fit factor is an important part of the decision system. It answers the question: Does the respirator provide enough protection considering the concentration of the contaminant in the air that the worker is exposed to?

The results of a qualitative or quantitative respirator fit test (QNFT) should be used to select specific types, makes, and models of negative-pressure respirators for use by individual respirator wearers. The respirator-fitting test should be carried out at least annually for each wearer who uses a negative-pressure respirator.

If the fit factor, as determined through an Occupational Safety and Health Administration (OSHA)-accepted QNFT protocol, is equal to or greater than 100 for tight-fitting half facepieces or 500 for tight-fitting full facepieces, then the QNFT has been passed with that respirator.

TRAINING PROGRAM

This portion of a respiratory protection program is most often left out but can least afford to be. Training brings the worker out of the dark. Training of employees includes the proper use of respirators, putting on and removing them, any limitations on their use, and their maintenance. Training answers a lot of wearer concerns and questions and reduces their scepticism. Wearers will believe the respirator is to their benefit when they participate in the program.

Training should be given to all people involved in the program, especially supervisors, the person issuing respirators, and the wearers/workers. Training should be carried out by a competent person and should include, as a minimum, the following items:

● Instruction in the nature of the hazard, whether acute, chronic, or both, and an honest appraisal of what may happen if the proper device is not used.
● An explanation of why engineering controls are not being applied or are not adequate and what type of effort is being made to reduce or eliminate the need for respirators.
● An explanation of why a particular type of respirator has been selected for a specific respiratory hazard.
● Instruction and training in actual use (especially a respiratory protective device for emergency use) and close and frequent supervision to ensure that it continues to be properly used.
● Instruction in how to recognize and cope with emergency situations.
● Instruction and training by competent instructors in the operation, limitations, and capabilities of the selected respirator(s) for the supervisor and wearers.
● Wearing instructions and training, including practical demonstrations, should be given to each respirator wearer and should cover:
  ● Donning, wearing, and removing the respirator.
  ● Adjusting the respirator so that it is properly fitted on the wearer and so that the respirator causes a minimum of discomfort to the wearer.
  ● Allowing the respirator wearer to wear the respirator in a safe atmosphere for an adequate period to ensure that the wearer is familiar with the operational characteristics of the respirator.
  ● Providing the respirator wearer with an opportunity to wear the respirator in a test atmosphere to demonstrate that the respirator provides protection to the wearer. A test atmosphere is any atmosphere in which the wearer can carry out activities simulating work movements, and respirator leakage or respirator malfunction can be detected by the wearer.
● An explanation of how to maintain and store respirators.

If a good training program does not exist, the most obvious effect will be that workers will be exposed to an air contaminant. The result of this oversight, of course, could be fatal when a toxic contaminant is used.

MAINTENANCE PROCEDURES

Each respirator used in the workplace should be properly maintained so that it retains its original effectiveness. It is recommended that this portion of the respiratory protection program include procedures for the following:

- Cleaning and sanitizing
- Inspection (After being cleaned and sanitized, each respirator should be inspected and tested to determine if it is in proper working condition or if it needs repair or removal from service.)
- Repair and testing (When inspection indicates the respirator needs repair, steps must be followed to ensure that proper repair and tests are carried out to confirm that the respirator is fully functional.)
- Storage
- Record keeping of maintenance, inspection, and testing procedures

MEDICAL SURVEILLANCE

As necessary, biological monitoring of employees should be performed at regular intervals. Initially, the capability of the worker to wear a respirator must be determined through pulmonary-function testing. Periodically, usually annually, a physical examination should be performed.

PROGRAM EVALUATION

A program should never be stagnant. It should be evaluated annually and its procedures updated based on criticism and suggestions. The evaluation should consider the following:

- *Worker acceptance.* The workers should be asked for comments on the respirator they are wearing regarding comfort, resistance to breathing, interference with vision, interference with communications, and confidence in respirator protection. Bringing the wearer into the decision-making process will provide greater support to the program.
- *Inspections.* Frequent inspections should be conducted to ensure that each step in the program is being followed.
- *Appraisal of protection afforded.* Results of biological testing and results of area monitoring will indicate the degree of effectiveness of the program.

The findings of the respirator evaluation should be documented and concrete target dates listed to implement corrective action.

CONCLUSION

In summary, in order to comply with the OSHA standard and other countries' occupational health and safety legislation and to provide the worker with adequate respiratory protection, a respiratory protection program must be established. This chapter summarizes the necessary components of this program. Some facilities may require an extensive program, whereas others may need only a minimum program. Manufacturing plants should require only one respirator program, whereas pilot plants or laboratories may require several, depending on the administration involved. Each facility should determine its current respirator usage and projected needs. If it is determined that respirators are not required at a facility, then a program is not necessary. A program coordinator should be selected in the plant to oversee the in-house development of the basic program.

Initially, setting up an effective respiratory protective program will require a lot of time and energy, but once established it will provide the company and employees with a concise and accurate step in the direction of better health and safety. Table 7.1 is a sample respiratory protection program that can be used as a guideline in developing your program.

Table 7.1   Respiratory protection program written format

PROGRAM	INSTRUCTIONS
I. *Introduction* In the control of those occupational diseases caused by breathing air contaminated with gases or aerosols, the primary objective is to prevent harmful exposures. This is accomplished as far as feasible by accepted engineering control measures (e.g., general and local ventilation, enclosure or isolation, and substitution of less hazardous processes or materials). When effective engineering controls are not feasible, or while they are being instituted, appropriate respirators may be required.	

Table 7.1 *(Continued)*

PROGRAM	INSTRUCTIONS
II. *Purpose and Scope* The practices and procedures described here constitute the program under which respirators are effectively utilized at_____.	Underlined statements or blanks are to be filled in by the plant to fit the program to the specific needs of the plant or area where respirators are used. The name of the plant or area goes here.

III. *Responsibility*
A._____is the respirator program coordinator. He is responsible for:
  1. Provision of appropriate respirators.
  2. Implementing training and instruction programs.
  3. Administering the overall program.
B. *Supervisory Personnel* are responsible for:
  1. Ensuring that respirators are available as needed.
  2. Ensuring that employees wear respirators as required.
  3. Inspection of respirators on a regular schedule.
	The title of one individual in the plant should be named here as overall coordinator of the respirator program. This person might be the plant engineer, the plant safety coordinator, plant superintendent, or a process engineer.

C. The *employee* is responsible for:
  1. Using the respirator supplied to him/her in accordance with instructions and training.
  2. Cleaning, disinfecting, inspecting and storing his/her respirator.
  3. Reporting a respirator malfunction to supervision.
	The employee's responsibility here is important. Effective training and supervision will be essential. Some plants with large respirator programs may want to assign a specific individual to clean and inspect all respirators. If this is the case, this paragraph may be rewritten to reflect the actual situation. The employee should, however, always inspect respirator before use.

D. The_____is responsible for: 1. Ensuring that the _____ standards include the requirement for respirator use when necessary.	Fill in the first blank in this paragraph with the title of the person who will review the process description used in your plant and determine at what point the respiratory protection equipment should be utilized. (This could be the process engineer, respirator program coordinator, etc.)

Table 7.1   *(continued)*

PROGRAM	INSTRUCTIONS
	The second blank in this paragraph refers to the name given to the process description or operating standard used in your plant. Please fill in the appropriate term.

E. *Industrial Hygiene Services* is responsible for:
  1. Technical assistance in determining the need for respirators and in the selection of appropriate types.
  2. Providing surveillance of work area conditions.
  3. Periodically evaluating the respirator program.
  4. Providing educational materials to be used in employee training.
	Surveillance of the work area conditions is essential to an effective respirator program. Air sampling and industrial hygiene surveys are needed to determine the proper application and safety of respirator use.

F. *Safety* is responsible for: 1. Technical assistance in the selection of respirators for use in emergency reentry situations and the training and instructions necessary for their use.	The coordinator of the respirator program should discuss with his Safety Engineer the need for emergency respirators. The need will depend on the type of potential hazards within the plant.
IV. *Respirator Selection* Respirators are selected by the respirator program coordinator and Industrial Hygiene Services. This choice is based on the physical, chemical, and physiological properties of the air contaminant and on the concentration likely to be encountered. The quality of fit and the nature of the work being done also affect the choice of respirators. The capability of the respirators chosen is determined from appropriate governmental approvals, manufacturers' tests, and plant experience with the respirators.	

Table 7.1   *(continued)*

PROGRAM	INSTRUCTIONS
V. *Distribution* Respirators are issued to individuals whenever possible. Each respirator that is individually assigned is identified in a way that does not interfere with its performance.	
VI. *Inspection and Maintenance* Respirators are properly maintained to retain their original effectiveness by: periodic inspection, repair, cleaning, and proper storage.	
A. *Inspection*	
1. All respirators are inspected routinely by the user before and after each use and after cleaning, to check condition of the face piece, headbands, valves, and hoses; and canister, filter, or cartridge fit.	
2. The foreman or supervisor inspects all respirators at least once per month.	This includes those respirators that are not used routinely, as well as those that are. You may wish to establish a method of recording this inspection. If so, it should be indicated at this point in the program.
3. Respirators maintained for emergency use are tagged noting the date of inspection, and the initials of the person doing the inspection. A log indicating these inspections is maintained in the plant office.	These tags are available and can be supplied to you for this purpose; or, you can develop your own tag system.
B. *Maintenance* Respirators that do not pass inspections are replaced or repaired immediately. Repair of the respirator by the user is limited to changing canisters, cartridges, filters, and head straps. All other replacements or repairs are performed by_____,	The ''experienced person'' mentioned here should be a designated technician who is trained or experienced in the repair of all respirators used. If he has a title, you may wish to substitute it here.

Table 7.1 *(continued)*

PROGRAM	INSTRUCTIONS
an experienced person with parts designed for the respirator. No attempt is made to replace components, or make adjustments, modifications, or repairs beyond the manufacturer's recommendations.	

VII. *Cleaning*

Individually assigned respirators are cleaned and disinfected as often as necessary to ensure that proper protection is provided for the wearer. Respirators not individually assigned and those for emergency use are cleaned and disinfected after each use. The following procedure is used for cleaning and disinfecting respirators:

A. Filters, cartridges, or canisters are removed before washing the respirator and discarded as necessary.

B. Respirators are washed in a detergent solution, rinsed in clean water and allowed to dry in a clean area. A brush is used to scrub the respirator to remove adhering dirt.

VIII. *Storage*

PROGRAM	INSTRUCTIONS
After inspection, cleaning, and necessary repairs, respirators are stored to protect against dust, sunlight, heat, extreme heat, extreme cold, excessive moisture, or damaging chemicals.	It is recommended that respirators be stored in plastic bags or the original cartons and placed in specially designed cabinets or lockers with other protective equipment. Under most conditions, it is not advisable to store a respirator in a toolbox or in the open. Cartridges or canisters and masks equipped with these components should be sealed in plastic bags to preserve their effectiveness. Improper storage will usually result in reduced service life and added cost.

Table 7.1 *(continued)*

PROGRAM	INSTRUCTIONS
IX. *Training* Every employee who may have to wear a respirator is trained in the proper use of the respirator. Both the employee and his supervisor receive this training. This training includes:	Training program materials for respirator use will be made available by Industrial Hygiene Services. Training, and reference materials may be attached to the respirator program as additional appendices.

A. Description of the respirators
B. Intended use and limitations of the respirator
C. Proper wearing, adjustment, and testing for fit
D. Cleaning and storage methods
E. Inspection and maintenance procedures

This training is repeated as necessary, at least annually, to ensure that employees remain familiar with the proper use of respiratory protection. The training program is evaluated at least annually by the program coordinator to determine its continued effectiveness.

X. *Records*

PROGRAM	INSTRUCTIONS
The following records are maintained by: _____.	Record-keeping personnel are to be determined by the plant. Please fill in the title of the individual selected. The respirator program coordinator is suggested.

A. The number of types of respirators in use
B. A record of employee training programs
C. Inspection and maintenance reports
D. Medical certification that the employee is capable of wearing a respirator under his given work conditions

OSHA regulations require a medical determination that an individual required to use a respirator must be physically able to perform the work and use the equipment.

Signed:_____
　　　　Respirator Program
　　　　Administrator/Coordinator

Approved:_____
　　　　Plant Manager

　　　　_____
　　　　Industrial Hygiene Services

8

Case Studies and Industrial Research

In the first edition of this book, we provided you with the case studies of problems associated with industrial use of respirators based on our experiences.[1] In this edition, in addition to those case studies, we also provide the abstracts of research studies performed by others with appropriate references. In our case studies, first-hand information was obtained by actually visiting a variety of workplaces where respiratory hazards existed. We observed that the proper use of respirators is difficult to achieve because of the following factors:

- The lack of proper training programs, including information on limitations associated with respirator use
- The lack of proper inspection and administration
- The user being reluctant to accept a particular respirator because of additional stress that it may cause, including interference with his ability to see, his freedom to move, and his ability to communicate
- The lack of testing facilities on-site, resulting in incorrect selection and improper fitting of respirators
- The lack of a proper maintenance and care program, resulting in improper functioning and reduced efficiency of the respirator
- A lack of understanding of NIOSH approval criteria
- Confusion regarding the use of respirators in conjunction with engineering controls
- The problems associated with specific acronyms and terms used by experts

In our opinion, the most effective way to illustrate the most commonly occurring respirator misuses is by citing the case studies.

CASE STUDIES (OUR EXPERIENCE)

Lack of Training, Including Information on Limitations

A workman entered a sewer manhole located in a street just outside a brewery. After a few minutes, he felt tired and weak. He tried to climb out of the manhole, but collapsed. A worker in the brewery, seeing this, rushed over with a respirator. The workman's helper grabbed an immediately available respirator, put it on, and proceeded into the manhole. Within a few minutes, he too passed out. When the two men were finally rescued by the Fire Department, they were both dead.

This incident clearly illustrates the hazard of selecting improper respirators. In this case, carbon dioxide from the brewery had entered the sewer and displaced the air. The two men died from a deficiency of oxygen; because the respirator used by the rescuer was designed to remove ammonia from air, it could not supply oxygen for breathing.

The following recommendations were made to prevent this situation in the future:

- Do not permit entry into an enclosed space without first checking its oxygen content. Do not enter any atmosphere with less than 19.5% oxygen without appropriate respiratory protection. When there is a question concerning the oxygen content of the atmosphere, an oxygen-deficiency measurement should be made.
- Before entry, inform the individuals making the entry about what hazards will be encountered.
- If the area is oxygen-deficient, only a self-contained breathing apparatus (SCBA) should be used.
- The individual making the entry must wear a lifeline controlled by an individual outside the oxygen-deficient atmosphere. The person outside must also wear SCBA and maintain visual contact with the individual in the oxygen-deficient atmosphere at all times. A third individual, also wearing SCBA, must also be standing by to provide aid if necessary.
- A training program must be provided giving adequate instruction on the limitation of equipment.

Lack of Inspection and Administration

A workman who entered an underground fuel oil tank to check the amount of sludge at the bottom of the tank was wearing a World War II gas mask. Shortly after entering the tank, he passed out. A second man, also wearing a World War II gas mask, entered the tank to rescue his fellow worker. He attached a rope to the unconscious man and then collapsed face down in the oil and sludge, tearing the air hose from his mask in the fall. The first man was pulled from the tank and recovered quickly. Because of the position of the rescuer after collapsing, he aspirated considerable oil and sludge and died 12 hours after being pulled from the tank.

In this case, a high concentration of hydrocarbon vapors, sufficient to cause narcosis, was present in the tank. A World War II gas mask was originally designed to remove war gases from air, and it is not certain whether it is capable of removing hydrocarbons.

The following recommendations were made for the future:

- A standard procedure must be developed to include all information necessary for the safe use of respirators in routine or emergency situations.
- Possible emergency uses should be anticipated and recorded in writing.
- The correct respirator for each job must be specified by a qualified individual supervising the program.
- The rescuer or standby man must be equipped with a SCBA.
- Frequent, random inspections must be conducted by a trained individual to assure that respirators are properly selected, used, cleaned, and maintained.

Lack of Maintenance

During one of our surveys, a worker was found wearing a respirator with the exhalation valve jammed in the open position. Thus, dust was allowed to leak through the respirator. The obvious need for corrective maintenance was pointed out to the employee with the following recommendation:

- All respirators must be inspected for defects (including a leak check), cleaning and disinfecting, repair, and storage.
- All respirators must be inspected before and after each use, and on a routine basis.
- If the respirator is not used routinely and stored between uses, it must be inspected at least once every month and kept in good working condition.
- When inspecting a respirator, all connections must be tight and the face-piece, headbands, valves, and connecting tubes must be in proper working order.
- Whenever respirators are inspected, tested, and repaired, the work must be performed by trained individuals.
- Instruction should be provided to the trainee regarding how to determine if a respirator is in good operating condition.

Improper Selection of Respirator

The operation of a fully enclosed trichloroethylene vapor degreaser was restricted by parts falling into and covering the heating elements. The degreaser was shut down at 9 a.m. and by 11 a.m. the degreaser had been drained and a compressed air-line set up to blow out the enclosure. In addition, an exhaust fan at the discharge end of the degreaser was turned on.

At about 11:45 a.m., a maintenance worker climbed down a ladder into the degreaser to recover fallen parts. He was wearing a cartridge-type respirator. A second man was posted on a platform above the degreaser. The first man collapsed in the degreaser. The second man called for help and climbed down into the degreaser to rescue his coworker. He also collapsed. Before the event was finished, six other fellow workers had entered the degreaser, and three firemen were called to the scene. Everyone except one fireman was to some extent overcome by the solvent vapors. The firemen were wearing SCBA.

There was an inch or so of solvent left in the degreaser after draining, but the surface for evaporation is more important than the total volume of fluid. In this case, the air above the surface could be saturated with trichloroethylene (as high as 9.78%). Because of the heavy vapor, there would be little tendency for the air-moving system to remove vapors.

A cartridge-type respirator is suitable for not more than 0.1% vapor. A gas mask could be adequate for 2% vapor, but only an air-line respirator or self-contained air-supplied respirator would have been suitable in this case.

The following recommendations were made:

- Before entering an atmosphere, air samples must be taken to ensure the compatibility of the atmosphere with the use of an available respirator.
- Samples must be taken and analyzed by a qualified hygienist.
- The selection of a respirator must be made by an individual who is qualified to do so.
- The user must be made aware that the cartridge cannot be used in atmospheres containing concentrations greater than those stated on the cartridge.
- Standby men or rescuers must always use SCBA.

Reluctance to Use Respirator

Three men were pumping the contents from a tan-yard vat when the pump inlet plugged. One man entered the vat to unplug the line and passed out. A second man entered to recover the first victim and he also passed out. A third man entered to assist and felt dizzy, but he was able to crawl from the vat using a ladder. Within a few minutes, he had help. He warned a fourth man not to enter the tank, but the man entered anyway, and he also passed out. With further help, the three bodies were removed, but none of them could be resuscitated.

An investigation revealed that proper respirators were available on-site, but none of the men bothered to use them. The vat had not been in use for about four years. The sludge from the vat, when aerated, released up to 4 mg/liter air (2700 ppm) hydrogen sulfide. The men, by disturbing the sludge, released heavy concentrations of hydrogen sulfide.

Reluctance to Use Proper Respirator

In an auto-body shop, all painters engaged in the spray-painting of cars with polyurethane paints were using half-masks with charcoal filters. They were all found to be sensitized against isocyanates. Investigation revealed that, although supplied-air respirators were available, the painters did not like using them due to the inconvenience. They tended to use filtering devices, typically half-masks.

In the two cases cited previously, the following recommendations were made:

● All workers must be educated regarding the benefits of wearing respirators when required.
● The toxic materials as well as their expected concentration in each operation should be identified. Also, the possible effect on the body of long- or short-term exposure (if the respirator is not worn), and what protection the respirator will provide, should also be explained.
● The worker must be taught to read the label on the filter cartridge to be sure that the proper filter is being used for the contaminant present.
● Positive and firm enforcement of respirator usage must be followed.
● Records of employees' negligence should be kept.

Lack of Testing Facilities On-Site

In a plant producing chlorine gas, individual exposures to chlorine were usually less than 1 ppm (time-weighted average). Occasionally, spills would occur, and for these emergency situations, the employees were provided with half-face chemical-cartridge respirators for their escape. It was found that, although most employees effectively used their respirators for safe evacuation, a few employees still required medical treatment.

Investigations revealed that the major cause for the trouble was poor facepiece-to-face seal because of facial hair and facial features. Because there were no facilities available on-site for either qualitative or quantitative fit tests, it was virtually impossible to determine who needed better protection (i.e., full-face mask, rather than half-face mask).

The following suggestions were made:

● Each worker should be provided with a respirator, which will provide him with the same level of protection that his fellow employees are receiving.
● Qualitative fit tests must be done for each respirator/wearer combination to determine facepiece to face-seal leakage.
● Positive and negative fit tests should be performed by the wearer before entering a contaminated atmosphere.

- If the leakage is considerable, due to facial features, a full-face mask respirator must be used for a better seal.
- Each person who wears a respirator should be checked for fit at least annually by a qualified individual. Aging, gains or losses in weight, or any of several other factors can affect the fit of a particular mask, and may make necessary a change to another brand.

Misinterpretation of NIOSH Testing and Approval Criteria[1]

In this case, insulation containing asbestos was being removed from steel beams. Because the worksite was not permanent, local or general ventilation was not feasible. Respirators were the only means available to control the exposure to dust. A visit to the site revealed that the workers engaged in removing insulation were wearing nonapproved disposable dust respirators, and the tradesmen in the area were not wearing any respirators.

During our visit, it was observed that tradesmen not wearing respirators grabbed the first available respirator lying around the job site and put it on.

Discrepancy in NIOSH Approvals

In an asbestos removal job, the contractor had supplied his workers with the type of respirators listed under NIOSH TC-21C type S.[2] During our visit, we recommended that instead of type S, type A should be used to better protect the employees. The contractor objected to our recommendation because, in his opinion, there was no difference between type A and type S. The NIOSH *Certified Equipment List*[2] describes type A as "designed as respiratory protection against asbestos-containing dusts and mists," and type S as "designed as respiratory protection against pneumoconiosis and fibrosis-producing dusts, or dusts and mists." Considering the wording used to specify each protection type, selection of type A is the obvious choice of worker protection for asbestos environments; however, because asbestos is a pneumoconiosis-producing dust, selection of type S also appears to be appropriate according to NIOSH.

Unfortunately, the published lists of certified equipment by NIOSH do not clarify these apparent discrepancies, and the respirator manufacturers tend to highlight these vague areas in their advertising. For example, the container of one type of single-use respirator states: "NIOSH/MSHA approved for dusts and mists (including lead and asbestos)." The respirator in question has an approval under the S category. It is clear that the manufacturer has designed the package label to take the broader meaning of pneumoconiosis-producing dusts (i.e., including asbestos). The person selecting a respirator would not be able to verify the manufacturer's claim(s) using the current NIOSH *Certified Equipment List*.

An experienced and trained person with industrial hygiene experience will usually opt for more stringent protection and select type A; however, an ordinary respirator selector would tend to read the label and select a respirator on this basis without further verifying the details. A compliance officer from the government would also invariably consider type S as complying with the requirement of controlling workers' exposure to asbestos.

The following recommendations were made:

- The contractor should provide respirators that are appropriate, under the circumstances, for the type and concentration of airborne asbestos.
- Ideally, all dust exposures should be thoroughly evaluated by air sampling before respiratory protection is prescribed. In the construction industry, this is not always possible because the worksite changes so often that by the time the sampling is done and results are available, the job is finished. In situations such as this, when adequate information is not readily available, it must be assumed that the greatest hazard possible is present, and consequently, the use of type A respirators listed on the NIOSH *Certified Equipment List* should be made mandatory.
- A respirator that is not approved by NIOSH or an equivalent agency is considered to be nonapproved. It should not be used for protection against asbestos under any circumstances.
- When it is not possible to provide adequate engineering controls for dust-producing operations, it is imperative that any persons coming into the area wear approved respirators. The fact that people (i.e., other tradesmen, supervisors) are not directly involved in the production of dust or may only be in the area for a short period does not absolve them from the need to be protected from the dust by wearing a proper respirator.
- Respirators lying around the worksite gathering dust should never be used by any person in the area for the obvious reason that one may inhale an excessive amount of dust in the first breath.

Engineering Controls versus Respirators

In one company, abrasive blasting using silica sand was used to clean metal parts in a booth ventilated according to the American Conference of Governmental Industrial Hygienists (ACGIH) *Ventilation Manual* criteria.[3] Both the blaster and his helper were wearing air-purifying respirators. When asked to change to air-supplied respirators, the employer maintained that ventilation had been provided to control the dust being produced and the respirators had been provided only as an extra precautionary measure. It was therefore unreasonable to ask for air-supplied respirators.

In order to convince the employer, several eight-hour personal samples for silica were taken. The results were 10 to 50 times the standard of $0.1\,\text{mg/m}^3$.

The following recommendation was made:

- For abrasive blasting operations involving silica sand, ventilation in the blasting booth is provided merely to keep the dust down for visibility purposes.

Hence, air-supplied respirators must be used to protect the workers from the siliceous dust generated.

Extra Protection for Paint Sprayers

In an assembly plant for heavy-duty trucks, spray-painting was done using compressed air-spray guns in five spray booths. Paints containing lead were used in all the spray booths. In one of these booths, which happened to be the largest, truck cabs were spray-painted by two painters. The booth was equipped with a down-draft, waterwash-type ventilation system, as well as air curtains at the entry/exit openings of the booth. The results of ventilation measurements proved to be in accordance with the recommended design criteria for spray-paint booths outlined in the ACGIH *Ventilation Manual.*[3] No respirators were deemed necessary for the sprayers.

During routine monitoring of employees' exposures, it was discovered that the lead-air monitoring results were above the short-term exposure limit (STEL). The employer was asked to provide adequate respirators to the sprayers. The sprayers refused to wear them, alleging that "a properly designed ventilation spray-booth should not require use of respirators by spray painters." The following counter-argument persuaded the sprayers to wear the proper respirators.

In large paint booths, too many variables can affect the airborne concentrations during painting, so it is not safe to rely on ventilation alone. Differences in paint viscosity and compressed air pressure of the spray gun may increase the amount of paint deflected into the air due to *bounceback*, or the painter may need to stand upstream of the exhaust ventilation in order to paint a certain area of the truck. When there is more than one sprayer in the booth, spraying onto the same object from different directions may cause excessive aerosol generation into another's breathing zone before ventilation takes effect. Hence, a respirator is the only means of assuring the painter of the lowest practical exposure.

The following recommendations were made:

- In addition to the ventilation provided for the large paint-booths, the sprayers should be required to wear approved respirators, even if the ventilation meets the recommended criteria.
- Sprayers' exposures must be regularly monitored to determine the efficiency of respirators being used.

Unnecessary Use of Respirators

A large company was involved in dumping, mixing, and bagging a powdered product that contained a small percentage of asbestos and silica. A large amount of money was spent to install adequate local mechanical exhaust at each dust-producing operation. The company was concerned about the hazards of exposure to dust and, thus, as an extra measure of protection, tried to force all workers to wear respirators even though adequate ventilation was provided. The employees complained to a government authority that they were being subjected to unnecessary stress.

An investigation revealed that the ventilation provided was more than adequate to protect the workers from asbestos and silica exposure. The results of a sampling were one-fifth to one-tenth the recommended standards. The following recommendations were made:

- The purpose of providing engineering controls should be to protect the worker from respiratory hazards and thus make wearing respirators unnecessary.
- In the event that engineering methods of control are not adequate, respirators should be worn until improvements are made.
- As a last resort, if controls are not adequate and further improvements are not feasible, then the use of respirators may also be necessary.

Problems Associated with Air-Line Respirators[1]

In a plant producing brake lining, a significant exposure to asbestos was occurring at the operation where dry mix containing asbestos was hand-scooped from a tote box into a preform press hopper. Although the hopper was ventilated in a limited way, the tote box was unventilated, and considerable dust was produced each time the operator transferred a portion of the mix. Although work practices definitely affected the worker's exposure, it was found that no matter how carefully the operator worked, his exposure was always above the legal standard.

As an interim measure, until improvements in ventilation and possible process changes were implemented, the operator was required to wear a respirator. The company felt that a continuous-airflow helmet would give adequate protection.

Because two motor skids were used to move metal tote boxes into and out of the area, it was decided that the air-line hose could not be allowed to drag along the floor. Thus, a clothesline was installed above the hoppers and the air-line was coiled around this clothesline to prevent it from contacting the floor.

While this system seemed good in theory, a visit to the plant revealed several problems that were preventing the operator from receiving the protection he theoretically should have:

1. The air-line became badly tangled. This resulted in the operators having to tug and pull on the line. Thus, the respirator was a hindrance to him and, occasionally, the operator would remove the helmet.
2. Because of tangling and pulling, the air-line was worn in places and minor leaks resulted. This reduced the flow of air to the helmet and subsequently reduced respiratory protection.
3. Although the helmets were provided with clear visors, the visors were often left flipped up.

While continuous-flow air-line helmets can offer a high degree of protection, the abuses observed prevented the operators from receiving adequate protection. The air-line hose was too much of an inconvenience because it had to be kept off of the floor. As a result of the abuses and limitations of the air-line respirator in this situation, the company switched over to a powered air-purifying respirator, which provided adequate protection.

The following recommendations were made:

- Air-line respirators suffer from serious limitations because of the long air hoses connected to them; hence, before selecting an air-line respirator, other alternatives must be considered. It is possible that the air-line respirators provide too much protection for the job.
- The airflow inside the helmet must be regularly checked to detect reduction in airflow resulting from kinks or other obstructions in the air-supply line.

Problems Associated with Facial Hair

A bearded firefighter using a demand-type SCBA was asked by his employer to shave off his beard in order to provide a good face-seal. The firefighter complained to the Human Rights Commission alleging that it was an invasion of his personal and religious rights to be told that he could not have facial hair.

A ruling was made that the employer was right in asking the employee to remove his facial hair for the purpose of safety. Our recommendations were that:

- If a firefighter objects to removing his facial hair due to personal or religious reasons, then only pressure-demand SCBA should be provided to protect him from toxic contaminants because a demand-type SCBA maintains negative pressure in the facepiece, thus increasing the risk of leakage.
- Although the pressure-demand SCBA will likely give adequate protection, the service-life of an air tank may be reduced from 30 minutes to 20 to 25 minutes because of outward leaks caused by a beard.

OTHER PUBLISHED RESEARCH AND CASE STUDIES[4]

Fatality Assessment and Control Evaluation Report: Assistant Manager at Ice Rink Asphyxiated by an Oxygen-Deficient Atmosphere in Alaska

Source: Ontario Disease Surveillance Report 5(8):78–80, February 24, 1984.

The case of a 24-year-old male assistant manager for a shopping mall ice skating rink who was asphyxiated inside a compressor room while attempting to shut off a refrigerant gas leak was examined. The gas was chlorodifluoromethane-22 (CFC-22). The employer was the owner of a 170-store indoor shopping mall, which included a swimming pool and an ice skating rink. There was no safety policy or program in place. The refrigeration system had a long history of leaks. A large leak had been plugged in a makeshift fashion. A maintenance worker performing a routine check observed refrigerant oil oozing from under the doors to the compressor room. The victim, the maintenance supervisor, and a maintenance worker entered the compressor room through self-closing doors. The victim was wearing a cartridge-type respirator, which was inadequate in an oxygen-deficient atmosphere. A coworker called 911 after entering the compressor room and seeing two of the workers lying on the floor. The maintenance worker and supervisor were rescued by an emergency medical service team. The victim was not in plain sight, and so remained in the room a longer period before being removed. He died of asphyxiation by oxygen displacement. Two swimmers in the pool and rescue personnel were also affected by CFC-22 vapor, which had spread into adjacent areas. It was recommended that workers be adequately protected from recognized hazards by installing appropriate engineering controls; that a maintenance program be developed; and that a safety program be designed.

Measuring Performance of a Half-Mask Respirator in a Styrene Environment

Source: R.A. Weber and H.E. Mullins, *American Industrial Hygiene Association Journal* 61(3):415–421, 2000.

A workplace protection factor (WPF) study was conducted with a half-mask air-purifying respirator during fiberglass boat production. Styrene was the measured analyte, and the geometric mean WPF found was 39.7. Analytical detection limits, sample contamination, and pulmonary elimination from previous exposures or from skin absorption were identified as important considerations that can bias the WPFs measured. There were significant differences in the mean concentrations found inside the respirator when analyzed by time period. An increase in the concentration found inside the facepiece cavity and a decrease in the WPF over time was found for people with three or four measurements. This indicates either a change in performance of the respirator over time or a bias from low-level exposures during the day or skin absorption.

OSHA Citation for Deficiencies in Respirator Training and Use and Improper Storage of Respirators

Source: OSHA Website, www.osha.gov/media/oshanews/Sept98/osha367.html, News Release 98-367, September 3, 1998.

OSHA cited Inland Waters Pollution Control, Inc., of Johnston, Rhode Island, for alleged Repeat and Serious violations of the Occupational Safety and Health Act during a complaint inspection conducted between March 2 and June 19, 1995. Approximately 24 workers are employed at the Detroit, Michigan-based hazardous waste cleanup firm's Rhode Island facility, which is located at 275 Scituate Avenue in Johnston. This firm is engaged in the removal, disposal, or cleaning up of hazardous waste and hazardous materials at various locations. Employees performing this work are often required to use respirators and to enter and work in confined spaces, such as storage tanks and underground pipes, which may contain toxic or oxygen-deficient atmospheres. This inspection identified deficiencies in respirator training and use and improper storage of respirators among other violations.

Three of the seven alleged Serious violations, with $30,000 in penalties proposed, were: (1) failure to establish standard operating procedures for respirators (including procedures governing selection, training, fitting, cleaning, drying, inspecting, repairing, storing, and emergency use of respirators, medical surveillance and evaluation procedures); (2) respirator users not properly instructed in respirator use and maintenance; and (3) self-contained breathing apparatus (SCBA) respirators not inspected monthly and one SCBA was stored for use without its air cylinder being fully charged, SCBA inspection records not kept, respirators and face pieces stored in wet, dirty, unprotected or haphazard conditions.

OSHA Citation for Failure to Train and Instruct Employees in the Use and Limitations of Respirators

Source: OSHA Website, www.osha.gov/media/oshanews/Sept98/osha367.html, News Release 98-367, September 3, 1998.

OSHA cited Anchor Glass Container of Dayville, Connecticut, for a total of 32 alleged Serious and Other than Serious violations of the Occupational Safety and Health Act and proposed penalties totaling $61,000. The alleged violations were discovered during a pair of safety and health inspections conducted at the glass-manufacturing facility. Approximately 385 workers are employed at the plant, which is located at 581 Hartford Turnpike in Dayville. The inspections uncovered a variety of safety and health hazards basic to this type of manufacturing environment, including the selection and use of respirators, employee training, and protections for employees who work within confined spaces. The combined inspections resulted in 25 citations and $61,000 in proposed penalties. Among the alleged Serious hazards were (1) failure to select respirators on the basis of the hazards to which employees were exposed; (2) failure to train and instruct employees in the use and limitations of respirators; (3) failure to ensure respirators were stored in a clean and sanitary location; (4) failure to perform respirator fit-testing; and (5) employees with beards allowed to wear respirators.

Occupational Health Risks Associated with the Use of Irritant Smoke for Quantitative Fit Testing of Respirators

Source: S.W. Lenhart and G.E. Burroughs, *Applied Occupational Environmental Hygiene*, September 1993.

The National Institute for Occupational Safety and Health (NIOSH) conducted a health hazard investigation (HHE) in response to a request from a fire chief of the municipal fire department. The HHE request was received after four fire-fighters reported experiencing either skin irritation or eye irritation as a result of qualitative fit tests using irritant smoke. One of these firefighters had eye irritation severe enough to require treatment at a hospital. The issues addressed in this case study are (1) evaluation of the health risks associated with the use of irritant smoke for qualitatively fit testing respirator face pieces, and (2) recommendation of an alternative method that should be used for fit testing the face pieces of self-contained breathing apparatuses (SCBA).

Because firefighting activities often occur in "highly toxic atmospheres or those immediately dangerous to life or health," a quantitative fit test should be used by the fire department. Because the purpose of a fit test is to evaluate the seal between a face piece and its wearer's face, quantitative fit tests should not be conducted using an operating SCBA.

The specific circumstances that caused skin or eye irritation by firefighters during qualitative fit tests with irritant smoke are unknown. The positive pressure inside the facepiece of a properly functioning pressure-demand SCBA would be expected to be sufficient to prevent leakage of irritant smoke into even a poorly fitting facepiece. If leakage of irritant smoke is assumed to have occurred when a pressure-demand SCBA was being worn, then consideration must be given to the possibility that the regulator of the SCBA was not functioning properly. There-fore, all SCBA used by the fire department should be evaluated to ensure that each one is maintained in accordance with its manufacturer's recommendations.

Finally, the sampling results of this study provide evidence for the first time that high concentrations of hydrogen chloride are emitted from irritant smoke tubes in environments with low and high relative humidity and that exposure to the fume produced by these tubes should be considered an occupational health risk.

Deaths Involving Air-Line Respirators Connected to Inert Gas Sources

Source: J.B. Hudnall, A. Suruda, and D.L. Campbell, *American Industrial Hygiene Association Journal* 54(1):32–35, 1993.

During 1984 to 1988, the U.S. Occupational Safety and Health Administration (OSHA) investigated 10 incidents, with 11 fatalities, involving the inadvertent connection of air-line respirators to inert gas supplies. Seven deaths resulted from connecting an air-line respirator supply hose to a line that normally carried inert gas. Four deaths were caused by leakage or backfill of inert gas into a line that normally carried breathable air. Ten of the deaths were from nitrogen and one from argon. The circumstances of the 11 deaths indicated that coupling compatibility and supervisory oversight were major factors in the inappropriate supply of irrespirable gas to the respirators worn by these workers. Conscientious-ness among safety personnel to the hazards of asphyxiation by inert gas, and compliance with current OSHA regulations, the ANSI Z88.2 standard, and NIOSH respirator certification approval regulations would have prevented these fatalities.

Two Patients with Occupational Asthma Who Returned to Work with Dust Respirators

Source: Y. Obase, et al. *Occupational and Environmental Medicine* 57(1): 62–64, January 2000.

This study was conducted to assess the efficacy of dust respirators in preventing asthma attacks in patients with occupational asthma (asthma induced by buckwheat flour or wheat flour). The effect of the work environment was examined in two patients with occupational asthma with and without the use of a commercially available mask or a dust respirator. Pulmonary function tests were performed immediately before and after work and at one-hour intervals for 14 hours after returning to the hospital. It was found that in patient 1, environmental exposure resulted in no symptoms during and immediately after work, but coughing, wheezing, and dyspnea developed after six hours. Peak expiratory flow rate (PEFR) decreased by 44% seven hours after leaving the work environment, showing only a positive late asthmatic reaction (LAR). In patient 2, environmental exposure resulted in coughing and wheezing 10 minutes after initiation during bread making, and PEFR decreased by 39%. After seven hours, PEFR decreased by 34%. The environmental provocation tests in both patients were repeated after wearing a commercial respirator; this resulted in a complete suppression of LAR in patient 1 and of immediate asthmatic reaction (IAR) and LAR in patient 2. In conclusion, two patients with asthma induced by buckwheat flour or wheat flour in whom asthmatic attacks could be prevented with a dust respirator are reported. Dust respirators are effective in preventing asthma attacks induced by buckwheat flour and wheat flour.

An Evaluation of Respirator Maintenance Requirements

Source: L.M. Brosseau and K.Traubel, *American Industrial Hygiene Association Journal* 58(2):116–120, February 1997.

In an effort to evaluate the performance of negative-pressure air-purifying respirator inhalation and exhalation valves over time in actual-use situations, a survey was conducted to examine maintenance aspects of industrial respiratory protection programs. Thirty companies, primarily involved in hard-goods manufacturing or service industries, using negative air-purifying respirators were surveyed. The factors most often cited as contributing to the decision to use non-disposable air-purifying respirators were type of containment and exposure level; cost was not a major determining factor. Most respondents were using the respirators for protection from hazardous aerosols. Ninety-three percent of the responding companies reported having formal respiratory protection programs that included written manuals; 93% reported that the respirators were inspected; 96% reported that the respirators were cleaned and sanitized; 79% reported that maintenance was performed; and 79% reported that inhalation and exhalation valves were replaced. Employees were responsible for inspecting the respirators in 96% and for performing maintenance and replacing valves in 73% of companies requiring such activities. Most respondents indicated that the service-life of a respirator was between three and four years. Based on these results, the authors suggest that all respirator parts should be inspected before and after each use, that replacement parts should be available on site, that respirators should be cleaned regularly, and that training sessions should include hands-on practice with respirators and their maintenance.

NIOSH Warns Workers about Explosive Respirator Cylinders

Source: NIOSH 93-127, August 1993.

This update warned workers about the possibility of respirator gas cylinder explosions during refilling with compressed air. The case in point involved a 47-year-old firefighter who was killed when the neck portion of the cylinder separated and struck him in the upper chest and neck. Several other incidents were on record reporting the explosion of DOT-E 7235 4500 PSI cylinders. In October 1985, NIOSH and the U.S. Department of Transportation began requiring that these cylinders be retrofitted with a steel reinforcing ring. The cylinder involved in this specific fatal accident had not been retrofitted and was in service beyond its maximum 15-year service-life. It was estimated that as many as 8000 of these cylinders may remain in service without the required retrofit. NIOSH urges that all compressed gas cylinders be examined, and that any of this type be removed from service if they have not been retrofitted or if they have exceeded the 15-year period of service. The last hydrostatic retest date stamped on the neck should be identified and the cylinder removed from service if the date is more than three years old. All compressed gas cylinders should be treated with caution.

Alterations in Physiological and Perceptual Variables during Exhaustive Endurance Work while Wearing a Pressure-Demand Respirator

Source: J.R. Wilson, et al. *American Industrial Hygiene Association Journal* 50(3):139–146, March 1989.

The physiologic and perceptual effects of respirator wear during exercise were investigated in order to more clearly define the problems associated with long-term work and respirator wear. Subjects included 19 firefighters experienced in the use of respirators and 19 inexperienced individuals. Both physiologic and perceptual responses were evaluated during endurance walks to exhaustion both with and without the respirator. Subjects were asked to inform testers when they felt they could continue for only five more minutes. Breath-by-breath metabolic and respiratory data were collected on each subject during these final five minutes before exhaustion. Use of the respirator resulted in a significant increase in the physiologic effort of breathing to overcome the added resistances. The increased effort of breathing increased the perception of exercise intensity and resulted in an earlier termination of exercise, by an average of 13.5 minutes. The continued increase in the maximum pressure measured within the facepiece of the respirator during expiration over the first 30 minutes appeared indicative of an excessive ventilation response, which was confirmed by the significantly greater ventilation volume to aerobic capacity ratio. Individuals who continued for a significantly shorter period with the respirator were experiencing respiratory and psychological distress as evidenced by sharp increases in ratings of perceived exhaustion and breathing. The authors conclude that, given current respirator designs, the continuous use of these devices should be limited to a maximum of 30 minutes.

The Role of Personal Beliefs and Social Influences as Determinants of Respirator Use among Construction Painters

Source: M.C. White, et al. *Scandinavian Journal of Work, Environment & Health* 14(4):239–245, August 1988.

The relative importance of personal beliefs and social factors as potential determinants of intended respirator use among painters at construction sites was studied. The specific factors assessed included individual beliefs about the benefits and drawbacks of respirator use and the influence of perceived attitudes of employers and colleagues. The study population included a total of 246 regular and apprentice members of two local union affiliates in Fort Worth, Texas, and Memphis, Tennessee. Participants agreed to a health examination and completed a self-administered questionnaire detailing personal characteristics and their medical and occupational history. The frequency of past and intended future use of cartridge respirators was highest among the younger workers and was lowest among the heavy smokers. Statistically significant negative correlations were established between intended respirator use and beliefs that the respiratory would be uncomfortable, would get in the way, would cause difficulty breathing, and would make the painter feel closed in. A statistically significant negative correlation was also established between respirator use and the belief that others would think that the painter was foolish, and a significant positive correlation was determined between intended respirator use and the belief that the painter would be better able to produce healthy children. The authors conclude that personal beliefs influence a painter's intention to use a respirator and that beliefs concerning discomfort or inconvenience were more important than those concerning the health effects of respirator use.

Individual Work Performance During a 10-Hour Period of Respirator Wear

Source: A.T. Johnson, et al. *American Industrial Hygiene Association Journal* 58(5):345–353, 1997.

A study of the effects of fatigue and food deprivation on human task performance while wearing a respirator was conducted. Eleven volunteers, mean age 24.7 years, wore a U.S. Army M1741 full-facepiece air-purifying respirator for 10 hours. They performed a battery of cognitive, psychomotor, and motor tasks in blocks of three hours without food. In each block, the tasks were performed for 50 minutes of each hour with a 10-minute rest break. The tasks were arranged into one hour of light work, one hour of intermediate-level work, and 1 hour of heavy work. Three replicates of the three-hour block pattern were performed in a round-robin fashion. The light work tasks were grouped into computer tasks, hand/eye coordination tasks, and assembly tasks. The intermediate task consisted of 50 minutes of cycle ergometer exercise at 60 to 65% of the predicted maximum heart rate. The heavy task consisted of self-paced exercise on a ski machine for maximum distance performed at 70 to 75% of the predicted maximum heart rate. Subjects performed the same tasks on a day when they did not wear the respirator. The subjects completed the Profile of Mood States (POMS) as part of the light tasks and rated their perceived levels of exertion (RPE) and comfort level during the intermediate and heavy tasks. Five subjects performed the experiment again but were allowed ad-libitum access to high-carbohydrate foods during the break periods. No significant differences in any of the measured outcomes were seen between the food-deprived subjects and the subjects allowed food. Task performance scores remained relatively constant throughout the day, indicating that wearing the respirator had the same relative effect as time and fatigue progressed. The POMS scores and RPE and comfort ratings did not indicate psychological fatigue at the end of the day. The only task performance scores that were consistently below 90 were those associated with tasks where the respirator interfered with peripheral vision. The authors conclude that wearing a respirator appears to cause few, if any, performance and psychophysiologic effects. Food intake also does not appear to affect performance.

A Questionnaire Survey on the Use of Dust Respirators among Lead Workers in Small-Scale Companies

Source: Y. Aiba, et al. *Industrial Health* 33(1):35–42, 1995.

A survey of the use of dust respirators by lead-exposed workers in small companies was conducted. The study group was composed of 141 workers, mean age 46.8 years, employed at six Japanese companies, with fewer than 100 employees that manufactured lead pigments and stabilizers for polyvinylchloride plastics. They were routinely screened for lead overexposure. The workers completed a questionnaire regarding details of respirator use, including type of respirator, how long the filters were used, how the respirator was maintained, and training received in respirator use and maintenance. Of the workers, 22% wore unauthorized respirators, those not approved by regulatory authorities. Half the workers replaced the respirator filters within one week; however, some used their respirators for longer than a week before replacing the filter. Seventy-three percent used a knit cover despite warnings by the Ministry of Labor that this could compromise the face-seal. Half of the subjects reported that their health supervisor had instructed them on how to wear and store their respirators, but 45% of the workers improperly stored their respirators in the workplace or worker's locker rather than in the specified storage cabinet. The authors conclude that a certain proportion of workers handling lead products in small companies wear respirators incorrectly and maintain them in improper storage places as a result of their not being trained in proper respirator usage and storage.

Task Performance with Visual Acuity While Wearing a Respirator Mask

Source: A.T. Johnson, C.R. Dooly, and E.Y. Brown, *American Industrial Hygiene Association Journal* 55(9):818–822, 1994.

The effects of wearing a respirator combined with reduced visual acuity on performance during a console-monitoring task and hand/eye coordination test was evaluated in 24 men and 22 women. Wearing a respirator with no additional reductions in the visual acuity reduced performance in the console-monitoring test by only 1% and by 11% for the hand/eye coordination test; however, as vision was handicapped from approximately 20/20 to 20/100 by exchanging nontampered lenses for ones with decreasing clarity, performance degraded more quickly for the console-monitoring test than for the hand/eye coordination test. Indeed, the console-monitoring task, which was a saccadics test, proved to be the most sensitive to reductions in visual acuity. The authors attributed this sensitivity to the smaller letters on the video display terminal. Because the artificially produced vision handicap could occur under real working conditions (e.g., due to condensation inside the lens, deposition of particulates on the outside of the lens, liquid or solid material flowing down the lens, lens scratching from normal wear, and impaired vision requiring corrective eye wear), the authors stress the impact of even small respirator-related decreases in visual acuity on performance and productivity, especially during critical functions. In this experiment, a reduction in vision from 20/20 to 20/30 corresponded to a 20% reduction in correct answers during the console-monitoring test. Such an impact may have important economic implications and may call for the use of engineering controls instead of respirators in dealing with contaminants.

A Study on the Usage of Respirators among Granite Quarry Workers in Singapore

Source: S.E. Chia, *Singapore Medical Journal* 30(3):269–272, 1989.

The frequency and appropriateness of respirator usage among granite quarry workers in Singapore were investigated. All granite quarries with more than 40 male workers were included in the cohort of 201 workers. Workers were interviewed using a standard questionnaire. Respirators were checked to verify if they were the correct type for use against silica. Smoking histories were obtained. Results showed that 43.8% of the workers did not use respirators, that 10.4% used the wrong protective device, and that only 45.8% used the correct respirators. There was a significant difference in the usage of respirators among drillers and crusher attendants as compared with drivers and mechanics. Age, years of exposure to silica dust, and type of occupation were correlated with respirator usage. The latter increased with age and duration of exposure. Of the drillers and crusher attendants, 75% wore respirators all the time, while only about half of the mechanics and drivers did so. Reasons given for not wearing respirators all the time were "difficulties in communication" (53.6%), "hot and sweaty" (53.6%), "wanting to smoke" (35.7%), and "breathing difficulty" (25.0%). Smoking history did not appear to significantly influence respirator usage. The author concludes that workers need to be educated on the importance of wearing the correct type of respirators and on the need to wear them all the time.

Subjective Reactions to Wearing Disposable Respirators

Source: B.G. Jex Courter, *Proceedings of the Human Factors Society 35th Annual Meeting*, Vol. 1, San Francisco, California, September 2–6, 1991. Santa Monica, CA: Human Factors Society; 890–894.

A study of subjective reactions to wearing disposable respirator masks was conducted. The study group consisted of 20 physically fit volunteers, 10 males, who had not worn respirators before. They wore a simple pressed-fiber mask or two masks that contained multiple-filter paper layers, exhalation valves, and adjustable straps on two days while resting or while performing simulated work. Two masks were of the one-size-fits-all type, and the third was a size medium, the only size available. The masks were rated for mask and strap discomfort with and without jaw movement (used as a simulant for talking), mask leakage with and without jaw movement, occlusion of the visible field, perceived difficulty of inspiration and exhalation, face temperature, face perspiration, perceived overall exertion and exertion in the legs and chest, air quality, mask odors, perceived overall exertion according to the Borg scale, and estimated heart rate using a computerized monitor or a paper and pencil packet. The responses were compared to those obtained when the masks were not worn. Female subjects reported more face leakage with and without jaw movement when wearing all three masks. The two complex masks blocked a substantial portion of the visual field. Practice obtained with putting the masks on tended to counteract the leakage experienced by the females. The paper and pencil ratings were more reproducible than the computer ratings over the two days of testing. The author concludes that fit problems, particularly for smaller female faces, are serious problems with disposable respirators. Users with small face structures or those whose jobs require communication while performing hazardous tasks are not getting adequate respiratory protection.

Methylene Chloride Exposure in Furniture-Stripping Shops: Ventilation and Respirator Use Practices

Source: A.H. Hall and B.H. Rumack, *Journal of Occupational Medicine* 32(1):33–37, 2001.

A survey was undertaken of the ventilation systems and work practices observed at 21 small-scale furniture-stripping shops in metropolitan Denver, Colorado, following the reporting of four cases of acute methylene-chloride (poisoning, including two fatalities, between 1984 and 1988), to the Rocky Mountain Poison and Drug Center. In each of these cases, the workers involved were using no respiratory equipment. In two cases, an open door or window provided the only ventilation. In one case, it was not known whether the operational ventilation system was being used at the time. In the other case, the local exhaust system was apparently inoperative. In six of the 21 shops, a dip tank was used. At these locations, respondents stated that a respirator was never worn at one shop and was at least sometimes used at the other five. At six of the other shops not using a dip tank, respirators were never used. None of the 21 respondents used any type of supplied-air respirator or self-contained breathing apparatus. In 10 of the 21 shops, the workers had experienced dizziness, headache, or nausea while stripping furniture. Ventilation systems were viewed as impractical or overly expensive to the operators of these small shops. The authors conclude that safety practices in small-scale furniture-stripping shops may be inadequate to protect workers from the fumes of methylene-chloride.

Workplace Protection Factors of HSE-Approved Negative-Pressure Full-Facepiece Dust Respirators during Asbestos Stripping: Preliminary Findings

Source: R.J. Willey, S.N. Tannahill, and M.H. Jackson, *Annals of Occupational Hygiene* 34(6):547–552, 1990.

Workplace protection factors (WPFs) of negative-pressure full-facepiece dust respirators approved by the Health and Safety Executive (HSE), United Kingdom, for asbestos removal were determined. Three types of respirators were evaluated. Ambient air samples were collected during removal of ceiling tiles containing asbestos, asbestos lagging material, and other asbestos-stripping operations by personal sampling pumps. Simultaneous in-mask air samples were collected on 13-millimeter-diameter 1.2-pore-size sampling filters from workers wearing the respirators. The samples were analyzed for asbestos. WPFs were calculated by dividing the concentration of asbestos in the ambient air by the concentration inside the mask. The WPFs for the three respirators ranged from 11 to 2090, 26 to 3493, and 17 to 500. The geometric mean WPFs were 200, 577, and 120, respectively. The percentage of WPFs above 900 for the respirators were 16, 40, and 0%, respectively. The WPFs increased linearly with increasing fiber concentration in the ambient air. The authors conclude that the measured WPFs differ considerably from data supplied by the HSE and the manufacturers. The reason for the WPFs increasing with increasing ambient air asbestos concentration is not known. It is considered important that the HSE and the respirator manufacturers realize that respirators do not perform as well in the workplace as they do under controlled laboratory conditions.

Employee Facial Hair versus Employer Respirator Policies

Source: G.L. Holt, *Applied Industrial Hygiene* 2(5):200–203, 1987.

Recent cases of controversy between employees and employers concerning facial hair and the wearing of respirators were reviewed. Facial hair, which comes between the seal of the respirator facepiece and the wearer's face, reduces the respirator efficiency and makes the protection offered less than that which the respirator has been designed to provide. Negative-pressure respirators can provide protection only with a proper seal. Not only have employees filed suits referring to race and sex discrimination, but also in relation to religious discrimination over the facial hair issue. One case was won by the employer, who demonstrated that the facial hair on the employee would prevent an adequate seal on his respirator. This company had also demonstrated good faith in trying, albeit unsuccessfully, to find another position within the company that would allow the employee to work without potential toxic gas exposure. The authors encouraged companies to set reasonable policies and to suggest a no-beard policy only in cases where such a ruling can be adequately supported. Employers who find it necessary to insist on a policy of no facial hair for their employees must be certain that the policy is justifiable and reasonable, properly worded, adequately posted, considerate of religious dictates, part of a well-managed respiratory program, and enforced consistently.

REFERENCES

1. Rajhans, G.S., and D.S.L. Blackwell. *Practical Guide to Respirator Usage in Industry*. Boston: Butterworth–Heinemann, 1985; 139–149.
2. Supplement to the NIOSH *Certified Equipment List*. DHHS (NIOSH) Publication No. 82-106, Cincinnati, OH: NIOSH, October 1981. (Current Publication No., 2001-139, September 20, 2001.)

3. *Industrial Ventilation—A Manual of Recommended Practice*, 24th ed. Cincinnati, OH: American Conference of Governmental Industrial Hygienists (ACGIH), 2001.
4. Abstracts of published research studies can be found in OSHLINETM^{TM} with NIOSHTIC-2. Hamilton, Canada: Canadian Centre for Occupational Health and Safety, August 2001.

Appendix

INDUSTRIAL RESPIRATORY PROTECTION WEBSITES

OSHA Respiratory Advisor site: www.oshaslc.gov/SLTC/respiratory_advisor/mainpage.html

NIOSH Selected Topics (Respirator) site: www.cdc.gov/niosh/respinfo.html

AIHA Respiratory Protection Committee: www.aiha1.org/Committees/RPC/homepage.htm

ANSI Standards Information: webstore.ansi.org/ansidocstore/default.asp

CSA International: www.csa-intl.org/onlinestore/GetCatalogDrillDown.asp?Parent:1020

CCOHS OSH Answers: www.ccohs.ca/oshanswers/prevention/ppe/respslct.html? print

British Standards Institute: bsonline.techindex.co.uk

Standards Australia: www.standards.com.au/catalogue

Index

42CFR84, 55, 57, 61, 75, 80, 89

Abrasives, 15, 19
Absorption, 3, 66, 75, 85, 88, 120, 155
ACGIH. *See* American Conference of Governmental Industrial Hygienists
Acid gas, 59, 67, 85
Action of toxicants, 8–18
Acute effects, xi, 47
Adhesives, 19
Administration, ix, 80, 105–7, 138, 145, 146–7
Administrator, 106, 110, 114, 131, 132, 135
Aerodynamic diameter, xi
Aerosol, xi, xii, xiii, xiv, 9, 11
Aging, 150
AIHA. *See* American Industrial Hygiene Association
Air pollution, xi
Air sampling, 38, 40, 41, 42, 140, 151
Air-line respirators, 74, 75, 78, 153–4, 157–8
Allergen, xi
Allergic reactions, xi, 18
Aluminosis, 15
Alveoli, 9, 10, 15, 17
American Conference of Governmental Industrial Hygienists, xi, xv, 11, 40, 41, 43, 44, 151, 152
American Industrial Hygiene Association, xi, 44
American National Standards Institute, xi, 68, 91, 111, 119, 158
Ammonia, 9, 15, 20, 21, 22, 23, 24, 25, 30, 34, 35, 59, 62, 63, 68
ANSI. *See* American National Standards Institute

Approved, xi, xii, 55, 74, 78, 79, 80, 95, 150, 151, 164
Asbestos, 1, 15, 19, 24, 25, 26, 29, 31, 32, 35, 82, 150, 151, 152, 153, 164–5
Asbestosis, 15, 120
Asphalt paving, 19
Assigned protection factor (APF), xi, 90, 91
Asthma, 117, 119, 120, 158
Automotive, 19

Bakeries, 19
Battery manufacture, 19
Berylliosis, 15
Biological monitoring, xi
Boiler making, 20
BOM. *See* Bureau of Mines
Breakthrough, 62
Breathing resistance, 55, 57, 58, 59, 117
Brewing, 20
British Standards, 78, 80–2, 111
British Standards Institution (BSI), 81, 82
BS. *See* British Standards
BS EN 136:1998, 81,102
BS EN 137:1993, 81, 102
BS EN 143:2000, 81, 102
BS EN 145:1998, 81, 102
BS EN 149:2001, 81, 102
BS EN 270:1995, 81, 102
BS EN 371:1992, 80, 102
BS EN 372:1992, 80, 102
BS EN 400:1993, 81, 102
BS EN 12941:1999, 81, 102
Bureau of Mines, 78,79
Business machines, 20
Byssionsis, 16